AVIATION IN MANCHESTER

A SHORT HISTORY

BRIAN R. ROBINSON

PUBLISHED BY THE MANCHESTER BRANCH
OF THE ROYAL AERONAUTICAL SOCIETY

Preface

It is now fifty years since the inaugural meeting that led to the formation of the Manchester Branch of the Royal Aeronautical Society. These years have seen an enormous growth and development of aviation activities in the Manchester area, from light aircraft such as the Avro Avian flying from the newly opened grass aerodrome at Woodford, to today when jet aircraft from Manchester Airport span the world. With the advance of technology the simple wood and fabric biplanes of the twenties have become the sophisticated high speed aircraft now produced in Manchester's factories.

To mark the fiftieth Anniversary of the Manchester Branch the Committee decided it should publish a booklet on the history of aviation in Manchester to show how the City has been involved with the growth and development of aviation, right back to the early days of ballooning. Decisions are easily taken, but the compilation of a booklet of this sort is a difficult and time consuming task. We were very fortunate. Brian Robinson, who is well known for his interest in aviation history, offered to undertake this task for us. We are very grateful to him for the preparation of this excellent and authoritative booklet.

We would also like to acknowledge with thanks the support we have been given by the following local companies associated with aviation:

Hawker Siddeley Aviation Limited

Fairey Engineering Limited

Middleton Sheet Metal Co Limited

Aircraft Tanks Limited

Baxter, Woodhouse & Taylor Limited

whose sponsorship has enabled us to proceed with this publication.

We hope that this booklet, in portraying our aviation heritage, will be of interest to all who read it and may inspire those who are today involved with Manchester aviation to work for a future as worthy as the past.

J.B. SCOTT-WILSON
Chairman
Royal Aeronautical Society
Manchester Branch

ISBN 0 9505515 0 3

Printed in Great Britain by A.B. Printers Limited, Leicester.

Introduction

The production of a definitive history of all aspects of aviation in relation to the area centred upon the City of Manchester would require a number of large volumes. I am conscious that a booklet of this size can only scratch the surface of the vast amount of information which exists and it has been necessary to leave out far more than has been included. I have set out to give an indication of the many ways in which efforts to produce and operate airborne vehicles have impinged on the population of an area which has been taken as approximating to that of the recently formed county of Greater Manchester.

Whilst by far the greatest degree of success in producing effective means of carrying man, his goods and his weapons by air has attended the use of heavier than air machines, it has seemed reasonable to include reference to aviation's kindred science of aerostation. Balloons provided the first effective form of aerial transport and, whilst the balloon and airship were once regarded as totally outmoded, the former has in recent years regained popularity for mainly sporting purposes and the latter has once again been suggested as a potentially viable load carrier.

I hope that publication of this booklet will inspire others — possibly more competent than myself — to undertake research into the various aspects of local aviation history and to make the results of their efforts available for the use and interest of present and future generations. A great deal of detailed information remains to be uncovered and I should be pleased to hear from anyone who has recollections, records or pictures relating to the earlier days of aviation in the area.

The records from which this booklet has been compiled have been collated over a long period and have owed much to a wide range of authors, to the assistance of local librarians and to very many friends and correspondents whom it would be impossible to list in the space available. To all of them I offer my thanks and, in particular, I must pay tribute to Jean Wyatt who has converted my appalling manuscript into readable form and to Eric Harlin without whose help I should have found it impossible to put the finished product together. Finally, I must set on record my massive debt to my wife, Olive, without whose encouragement, tolerance and support neither the research nor the writing could have been undertaken.

BRIAN R. ROBINSON
25 Cromwell Grove,
Manchester M19 3QD

January 1977

:Manchester Central Library

The Balloon Era

It would probably be idle to speculate as to the precise stage of human development when man first eyed the birds and yearned to achieve their mastery of an additional dimension and the freedom of their progress through the air. From the earliest days of legend and history have come stories which evidence the continuing search for means of achieving flight and as man developed both his prowess and his ambition he added to his desire to escape his earthbound state an association of flight with superhuman speed. Hence it is, perhaps, not surprising that, when the emerging needs of commerce brought a desire to reduce the time of travel between Britain's developing centres of habitation, the vehicles used were frequently accorded the attributes of a later era.

The first flying machines known to Manchester's early population were the flying stage wagons of Mr Howarth of Flixton which raced along the perilous roads of the area from 1744 and ten years later the 'Manchester Flying Coach' established communication between Manchester and London, the journey time being approximately four and a half days. By 1770 it was possible to board the 'London Flying Machine' of Samuel Tenant at the Royal Oak in Market Street Lane and arrive in London only two days later.

Whilst these flying machines, aided by the rapid improvement in the standard of road surfaces, reduced the time taken to travel between the major cities of the realm, little progress was made toward the production of a device which would enable men to achieve actual controlled flight from place to place. Amongst the devices planned and sometimes produced over the years had been helicopters, ornithopters, gliders and powered and man-powered aeroplanes but none had been successful. It has been suggested that some form of balloon may have been evolved in South America almost 2000 years ago but it was on the 4th June 1783 that an event in France provided the first authenticated step toward enabling man to rise from the ground and remain for an appreciable period floating above his fellows, albeit at the mercy of the winds and without directional control of his progression.

The men responsible for this great step forward were the brothers Montgolfier, proprietors of a paper manufacturing concern at Annonay, near Lyons who, after observing the lifting forces present over a fire, conducted experiments which led to a demonstration in the market place at Annonay when a fabric sphere lined with paper was filled with hot air and released to rise into the sky and float silently away. Not surprisingly word of this achievement soon reached Paris and on the 19th September, 1783 the Montgolfier brothers sent up a larger balloon from the capital, this time bearing a sheep, a cock and a duck, all of which survived to become the first living creatures to fly successfully by other than their own physical efforts and to point the way for the first men to make an aerial journey. On the 21st November, 1783 Monsieur Pilâtre de Rozier and the Marquis D'Arlandes rose from the Bois de Boulogne beneath a Montgolfier balloon and travelled serenely above the rooftops of Paris.

Whilst the Montgolfiers had been working towards de Rozier's epoch making ascent, a Parisian scientist, Jacques Charles had been considering the use of hydrogen as a lifting agent and on the 27th August, 1783 he sent up a small balloon filled with that gas. The ascent was a success, although the balloon was sadly ill-used by the peasantry on landing, and on the 1st December, 1783 Charles, accompanied by one of the brothers Robert who had made the envelope, achieved the first ascent of a man-lifting hydrogen balloon.

Despite the absence of today's world-wide information media, news of the successful French balloons did not take long to reach England and to inspire others to experiment with similar vehicles. On the 26th December, 1783 a gas filled balloon was inflated in the grounds of the Manchester Infirmary and released to float on the wind for a distance of 45 miles before coming to earth at Cromford in the Peak District of Derbyshire. Experiments in the area do not appear to have produced a local aeronaut, however, and whilst, in the following year, the Italian, Vincent Lunardi rose from London beneath a hydrogen balloon and the Scot, James Tytler became airborne at Edinburgh in the basket of a hot air balloon, Manchester had to wait until 1785 before the first man took to the air from its environs.

However, if no Mancunian had found satisfactory means of getting himself off the ground, some attention had been given to the application of the balloon to less responsible pastimes and by the 29th October, 1784 the release of 'lighted balloons' had 'become so frequent and alarming to the inhabitants' — and no doubt prejudicial to their crops and wooden buildings — that the Boroughreeve and Constables felt it necessary to publish notices prohibiting them along with 'squibs, crackers and rockets'.

The first man to fly over the Manchester area was James Sadler, the son of an Oxford pastrycook, who had become the first English aeronaut when he ascended with a hydrogen balloon from his native city on the 4th October, 1784. He came to Manchester conscious that a balloon ascent was not merely a scientific achievement but a spectacle for sight of which the citizenry could be induced to part with hard cash. The balloon to be used was advertised to be on exhibition at the Exchange from the 31st March, admission to be one shilling, a not inconsiderable amount in relation to the prices and values of the day, and obviously not designed to attract the labouring masses, although 'working people and servants' could be admitted for six pence. Initial advertisements indicated that the aeronaut would be a man called Harper who had bought a balloon from Sadler and made an ascent at Birmingham but, by the 3rd May, it was announced that James Sadler would arrive in Manchester to ascend 'on Old May Day next'. It may have been that Harper was, in effect, acting as a publicity and advanced booking agent as on the 5th May Sadler was engaged in ascending from Moulsey Hurst on the Thames but by Thursday, 12th May preparations had been completed and there appear to have been no plans for Harper to join Sadler in the basket. He appears to have remained on the scene and it is possible that his role was to some extent that of financial guarantor.

The site chosen for the ascent was the garden at the rear of a gentleman's residence in Long Millgate. The house, that of John Hawarth, still stands as the Manchester Arms Hotel, now entrapped between Victoria Station and the large buildings of the Co-operative Wholesale Society. Corporation Street, which passes the back of the premises, crosses the former garden. The names 'Balloon Street' and 'Garden Street' recall the events of 1785 and on the premises at the corner of Corporation Street and Balloon

Street a ceramic plaque erected by the City Council commemorates the first balloon ascent from the area.

A crowd of 5,000 people assembled early to witness the ascent. Many of them had purchased tickets for entry to the grounds at half a guinea and five shillings each but no doubt many more had preserved their funds and remained beyond the enclosed area. The balloon filled slowly as iron filings were introduced to a vat of sulphuric acid to produce the hydrogen and by 1.20 pm Sadler ascended to the accompaniment of loud cheering. He remained aloft for an hour and three quarters—although another account suggests the time may have been shorter — and finally landed without harm on a sand bed between Blacow Bridge and Radcliffe to pack up his balloon and return to Manchester the same evening.

A week later on the 19th May Sadler ascended again from the same site and on this occasion he deemed it prudent to reduce the entry fee to the grounds to a mere three shillings, considering, perhaps, that this would induce more people to pay to witness the spectacle of his ascent. On this occasion he found the elements a little less forgiving and, having been carried to a height of 13,000 feet, he suffered from the cold and experienced sharp pains in his ears before descending at Pontefract in a wind of sufficient velocity to cause the balloon to drag him for a considerable distance before escaping and finally coming to rest near Gainsborough. To add to the pain of the injuries received on landing, Sadler suffered the further mortification of finding that the people who retrieved the balloon were more attracted by its silk material than by the sum which he offered for its recovery.

The French Revolution and the political and military upheavals which followed in its wake probably did much to limit the ballooning which took place in the following decades and it was not until 1812 that Manchester saw another balloon, the aeronaut again being James Sadler who came to the town to ascend from the grounds of Davidson's Rope Walk at the rear of a foundry at the corner of Swan Street. Sadler set up his headquarters at 3 York Street and exhibited his balloon and car at premises in Mosley Street. This was the 55 foot diameter balloon in which he later made an unsuccessful attempt to cross the Irish Sea.

Thousands of people assembled in the vicinity of Shudehill to see Sadler lift off into a stiff breeze, waving his hat with one hand and no doubt maintaining a firm grip on the basket with the other. The balloon passed over Ashton and descended toward the Pennines where a carter on a moorland road fled in terror as it collided with his vehicle. Sadler flung out a grapnel which held briefly before the wind swept the balloon away and he finally came down in a wood six miles east of Sheffield.

The tradition of ballooning remained in the Sadler family and a great-great-grandson of the pioneer is reported to have made captive ascents from the Dover area in an observation balloon as late as July 1941 when he may well have found more than the elements to contend with in that embattled area. From 1814 James Sadler's son William Wyndham Sadler toured the country, making his final ascent from Bolton on the 29th September, 1824 when after filling his balloon with coal gas he was blown toward Blackburn and collided with a factory chimney at Foxhill Bank, near Oswaldtwistle, falling to his death from a height of 60 feet.

During the 1820's balloon ascents began to be a popular feature of galas, fairs and celebrations in Manchester and the surrounding towns.

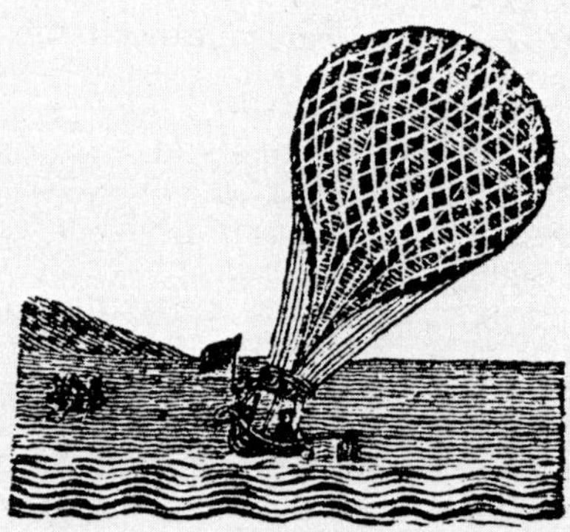

Mr. Sadler

Respectfully informs the Ladies and Gentlemen of Manchester and its vicinity, that he will make his TWENTY-THIRD ASCENSION from

Mr. Davenport's Rope-walk,

Behind Mess. PEEL, WILLIAMS, and Co.'s Iron Foundry, Shude-hill,

On Monday, June 29, 1812,

At One o'Clock.

Tickets of Admission, at 5s. and 3s. each, to see the Process of filling the Balloon, as well as its Ascension, may be had of Mr. Sadler, No. 3, York-street, and at the building, Mosley-street, where the Balloon and Magnificent Car is now exhibiting.—Admission, 1s. each.

To prevent the Confusion and delay incident to taking money at the entrance, Mr. Sadler respectfully suggests to the Public, the propriety of taking Tickets, prior to the Day of Ascension.

A Plan of the Ground and Accommodation may be seen where the Tickets are to be disposed of.

Manchester: Printed by W. Cowdroy, Gazette-Office.

A handbill produced to advertise one of Sadler's ascents from Manchester.
:Manchester Central Library

Before his fatal escapade from Bolton, William Sadler had ascended from the Cloth Hall on the present site of Salford's Victoria Bus Station to pass over Altrincham and descend at Monk's Heath near Knutsford. He was up again with two passengers on the 3rd May to celebrate the opening of Rochdale's Gasworks, landing at Rawcliffe near Bacup and on the 22nd September was again in the area to make a brief ascent from Wigan.

The aeronauts who became best known in the Manchester area in this period were undoubtedly George and Charles Green who were prolific exponents of exhibition ballooning and travelled widely throughout the country. With contemporary records frequently referring simply to 'Mr Green' and the added complication that either or both might ascend from a jointly prepared site, there is some difficulty, one hundred and fifty years later, in attributing particular ascents. The first local exhibitions which they arranged were from Bolton Gasworks and on the 11th September, 1826 the gasworks manager's daughter was taken for a flight which ended at Cinder Hill near Royton.

The Greens arranged two ascents from Ashton Gasworks in June 1827, the descents being at Thorne in Yorkshire and at Coombs Rocks near Charlesworth, before the first ascent from Stockport on the 18th June. On this occasion the balloon rose from Castle Yard at such a rate that Green and his passenger were able to see the sea within four minutes of ascending. By this time local coal gas supplies had superseded the use of hydrogen and it appears that the quality of the Stockport product was of a high standard.

Ballooning was still for a relatively limited few and the Greens made a practice of extending the numbers who could sample aerostation by making numerous tethered ascents to entertain the crowds before departure. At Macclesfield where he first ascended on the 25th June, 1827 Charles Green is reported to have made many limited ascents with three or four passengers each of whom paid five shillings for the privilege. Manchester saw both Greens in full flight on the 28th July of the same year when, after leaving Warrington, their balloon crossed the town before landing at Springhead.

George Green was at the Salford Cloth Hall on the 29th July, 1828 to make an ascent in aid of the Salford and Pendleton Dispensary and this occasion was notable for the first demonstration of a parachute in the Manchester area. As Green left the area, he released a live hen in a basket which floated down into Silver Street whilst the balloon, after floating over Levenshulme and Stockport, came down conveniently close to Poynton Hall.

One of the developments of the first half of the 19th Century was the emergence of the pleasure gardens where the masses and their alleged betters could view floral displays, attend concerts and sporting contests and, in some cases, marvel at collections of exotic animals and birds. The Collyhurst area has not, in recent years, been particularly notable for its country air and pleasant scenery but, as early as 1797, Robert Tinker had established his Coffee House and Tea Gardens between the Rochdale Road and the River Irk, subsequently expanding both his grounds and his enterprise until, in 1812, he adopted the name of the famous London Vauxhall Gardens. The name Vauxhall Street can still be found on old property in the area close to the modern public house which bears Tinker's name and includes in its decor scenes recalling the gardens and their attractions, not least of which were ascents by famous aeronauts of the period.

The first balloon to draw the populace to the Vauxhall Gardens was the crimson and gold creation of Charles Green who made his 79th ascent on the 10th August, 1827 with two passengers. The balloon was inflated with coal gas at the gas works in St George's Road and moved to the gardens by a team of men hauling on ropes whilst Green rode in the basket above the level of the houses. After being ceremonially presented with flags by the proprietor's daughter, Green and his passengers rose from the bowling green where they had moored for inspection by the crowds and drifted away over Ashton, Glossop and Sheffield to land four miles from Worksop. To keep the crowds at Collyhurst informed of their progress they released pigeons at intervals on their way.

Tinker apparently found the ascent a paying proposition as he engaged George Green for a similar event on the 10th September, 1828 when T.W. Strapps, a Salford baker who had been a passenger the previous year, again secured a place in the basket. Green visited the Gardens again on the 18th June, 1832 when Tinker organised a celebration to commemorate the passing of the Reform Bill and on the 12th September, 1836 no less a personage than the Duke of Brunswick attended to ascend with an aeronaut named Graham. On this occasion, however, the supply of gas proved inadequate and when, by early evening, inflation had not been completed, the restive crowd threw fireworks and attempted to set fire to the balloon. The reactions of the noble Duke are not recorded but police restored order and Graham eventually ascended alone late in the afternoon of the following day to reach the edge of Carrington Moss. A further ascent planned for a few days later does not appear to have taken place and Graham may well have decided to depart the area without again placing his faith in the dubious gas supply from St George's Road.

Control of the gardens passed from the Tinker family and, although other entertainments continued to attract the crowds, there were no more balloon ascents until October 1847, when Hampson was booked to display his 'Royal Irish Balloon'. By this time the practice — still in vogue — of giving each balloon a name had developed and the balloon bore the title *Erin-Go-Bragh* on its envelope but no title could compensate for the continuing deficiency of the local gas and, after abandoning his intention of taking a passenger, Hampson was content to struggle up and land not far away in Phillips Park. This was the last ascent from the site which vanished under a maze of streets for a century and more.

Whilst Robert Tinker was the pioneer of the large pleasure garden in the Manchester area, the Vauxhall Gardens found some rivals in the middle years of the nineteenth century, one of them to survive and prosper until the present day. In 1836 John Jennison bought a small inn named Belle Vue House in Gorton from a man called Walker who had already established public gardens close to the inn. Jennison included amongst his initial attractions a number of wild animals and laid the foundations of the present zoological collection. Although a man of considerable enterprise, Jennison did not engage his first aeronaut until the 2nd June, 1852 when as a Whit holiday attraction he invited Arthur Goulston, who had gained a growing popularity in the London area under the adopted title of Guiseppi Lunardini, to bring his balloon *La Normandie* to Belle Vue. Unable to accept due to a prior commitment at the Cremorne Gardens, 'Lunardini' arranged for his father, James Goulston, to travel north with a completely new balloon and John Jennison arranged to crown the display by ascending with him.

James Goulston was a carpet manufacturer and had made a large number of ascents as a patron of various aeronauts but had not previously taken command of a balloon. On the day fixed for the ascent he found that weather conditions were far from ideal and the supply of gas inadequate to enable a passenger to be carried. Although recommended by Jennison to defer the trip, Goulston was determined to fulfil the engagement and ascended during the evening into a strong wind which carried him rapidly over Newton Heath, between Bardsley and Oldham, and across the eastern end of Lees before the basket struck the ground at Stonebreaks Hill. In the midst of thunder and heavy rain he attempted to leap from the basket but his legs became entangled in the rigging and he was carried along until his head came into violent contact with a stone wall on which, according to a contemporary report, his blood and brains could still be seen some days later! Goulston was carried into the Three Crowns Inn at Austerlands but was found to be dead.

This tragedy did not deter the Jennison family from providing further airborne entertainment for their patrons and the next balloonist to appear there was Lieutenant Chambers, R.N. who brought an unusual 'wall-sided' balloon named *Victory* to make four ascents in August and September 1854, coming down in turn near to the White Lion at Withington, close to the Stamford Arms at Hattersley, in Denton and 'on a common at Moston.' Chambers is said to have used gas from Jennison's own works. Four years later one of the most famous British aeronauts Henry Coxwell made his 271st ascent from Belle Vue on the 7th July, 1858, taking as passenger in the basket of his balloon *Queen* a Mr Walker, clerk with Messrs Lingard & Vaughan, Solicitors of Stockport, to land at Gee Cross near Hyde. By this time the distance of travel had been subordinated to the provision of an attraction for the crowds paying to watch the ascent and during July and August Coxwell made a series of four short trips from Belle Vue to Timperley, Ashton Woodhouse, Mobberley and Rowarth.

The mere sight of a gas-filled globe rising into the sky was insufficient to persuade the increasingly demanding public to part with their money and for the last occasion on which the Jennisons were to engage a balloonist they brought to Gorton Hall's *Florence Nightingale* equipped with a hanging trapeze. Hall made two ascents on the 9th and 14th July, 1859 wearing fancy dress and performing on the trapeze as the balloon rose from the gardens. He came down intact at Coombs Rocks near Dinting on the first occasion and behind the Traveller's Call at Bredbury on the second, but his career ended in the same month when he was killed at Newcastle.

The third of the major pleasure grounds of the Manchester area to play host to the travelling aeronauts was the Pomona Gardens which grew from the early Cornbrook Strawberry Gardens on a site near Ordsall Hall and continued to attract crowds for varied events until 1888 when it disappeared to make way for Manchester Docks. At one time the Gardens boasted the largest ballroom in Britain and in their hey-day a crowd of 100,000 assembled in the grounds to hear Lord Salisbury and Lord Randolph Churchill not far from where the latter's great-grandson — an enthusiastic pilot — secured his seat in Parliament many years later.

As was the case at Belle Vue, aerostation at Pomona Gardens got off to a lively start, although the effects in this case did not prove fatal. On the

19th June, 1847 the American aeronaut, Gale, brought his *Prince of Wales* balloon which was advertised as having two auxiliary balloons and a duplicate car which would be lowered to enable Gale to descend by ladder from the main car and fire a Royal Salute. However, matters did not go entirely according to plan. After inflating the balloon from a street main, Gale climbed into the main car but the buoyancy was grossly inadequate and he discarded in turn his two passengers, his auxiliary car and a large quantity of ballast before careering across the lake, through the railings, over Chester Road and through a line of poplar trees, finally becoming airborne and eventually travelling via Hulme, St Peter's Square, Victoria Park and Heaton Chapel to Bredbury.

Gale did better a week later when his arrangements succeeded, including the discharge of various pyrotechnics and he crossed Stretford and Altrincham to land at Sinderland near Bowdon. In September he brought his new *Royal Cremorne Balloon* and made two ascents accompanied by his backer, James Goulston, landing at Kersal Moor and Hale Barns. Within five years both had been killed in ballooning accidents, Goulston at Stonebreaks Hill and Gale near Bordeaux in France.

The Pomona Gardens saw several other ascents but the next balloon to be displayed did not leave the ground and the intrepid Mrs Graham, who had reached the stage of entering the basket only to find the balloon unable to lift her, subsequently obtained redress in the Manchester County Court for the poor standard of gas supplied. Goddard made two ascents in his *Aurora* in 1860 and the final aeronaut to rise from Pomona Gardens was William Henry Adams who took off on three occasions in July and August 1865. The balloon bore the name *Garibaldi* and, whilst his first two descents at Middleton and Sale Moor appear to have been uneventful, Adams had the misfortune to land at Ancoats after his last ascent and to have the balloon badly damaged by a mob of Irishmen who took exception to its name.

Virtually all the aeronauts who ascended from Manchester came to the area from other parts of the country and there do not appear to be any records of any Mancunian balloonists in those early days. However, at least one man in Manchester did direct his attention to improving the science and safety of aerostation. Dr A. Crestadoro, who was Chief Librarian of the Manchester Free Library and a member of the Aeronautical Society, published a pamphlet on Air Locomotion and in 1868 took out a patent relating to a balloon which was described as 'the only fire balloon which seems free from the danger of conflagration.' A model of his metallic balloon was exhibited at the Society's Exhibition at the Crystal Palace in the same year.

Not long after Crestadoro's experiments had been made known to the nation's scientists his library and the day to day reading public of Manchester owed to the balloon a totally different contribution to their knowledge. In July 1870 the Franco-Prussian War broke out and the City of Metz was soon besieged. The only British War Correspondent in the city was George Robinson of the Manchester Guardian and, faced with the apparently total absence of ways of sending his despatches, Robinson hit upon the idea of making small fabric balloons, filling them with coal gas and releasing them to drift on the wind over the Prussian lines carrying messages for onward transmission by the finders. He also wrote messages on the sides of his balloons and, although many were lost, the Guardian eventually received

The Nassau Balloon in which Charles Green ascended on three occasions from Salford during 1837. *:Science Museum, London*

some of his reports.

Whilst balloon ascents from Manchester were mainly from places of regular public entertainment, the surrounding towns of lesser size saw balloons less frequently and the arenas from which they ascended were often selected primarily for their proximity to a source of gas supply. Charles Green with his 70,000 cubic foot *Royal Vauxhall* Balloon made an epic flight from London on the 7th-8th November, 1836, crossing the Channel and finally landing at Nassau in Germany. He brought the same balloon, by then re-named *Nassau*, to Salford the following year and made three ascents. The first of these provided an unpremeditated additional thrill for the onlookers when, with Green and a passenger aboard, the balloon was released prematurely and Green's assistant, Carpenter, was carried aloft hanging onto a single rope. Fortunately, the man was — to quote a contemporary report — 'bred to the sea' and climbed the rope safely to enjoy a view extending to the Welsh coast during the balloon's passage over Rochdale, Oldham, Ashton and Stockport. Admission to the Gasworks ground to witness the ascents had by now been reduced to one shilling and ladies and gentlemen wishing to accompany Mr Green could do so for ten guineas and twenty guineas respectively, the balloon car having seats for ten people. Two other ascents were made during Green's stay and passengers were carried to Winterset near Wakefield and to the field of an Offerton farmer, one Mr Fogg who immediately demanded £5 for damage done to the field. His demand was not met, Green pointing out somewhat reasonably that the potato crop could hardly have been damaged as it had already been lifted!

Green's balloon *Victoria* was displayed at Stalybridge Gasworks on the 23rd July, 1850 before ascending under the command of Isaac Watt Boulton and travelling north to Garstang and a month later Boulton took the same balloon from Ashton-under-Lyne Market Place eastward over Woodhead to a point near Gainsborough from which three pigeons were released and arrived back in Ashton with news of the flight. Another Green balloon *La Continent* appeared at Tommy Field, Oldham in May 1859 when George Green and a local broker named Woollacott crossed Manchester to come down on the farm of Edward Rogers at Davyhulme.

Castle Yard, Stockport, which had been the site of the town's first balloon ascent in 1827, was again used for the inflation of Goddard's balloon on the 30th September, 1859 in celebration of the opening of Vernon Park a year previously. The balloon crossed Hyde and Stalybridge and eventually landed on Greenfield Moor in a place sufficiently remote to cause Goddard and his companion Peter Fosbrooke to roam the moors for three hours before obtaining help. Fosbrooke, an officer of the Earl of Chester's Yeomanry, does not appear to have been put out by the experience as, in September 1860 he secured a place in the basket of Lythgoe's *Prince of Wales* balloon to travel from Stockport to Pickford's Farm near Bramhall Station. The crowds had good value on this occasion as several small balloons were sent off during the afternoon and 'numerous Montgolfiers were despatched from Churchgate.'

Henry Coxwell brought his *Mammoth* to Bolton for the inauguration of a statue of Samuel Crompton after spending several months on crutches following a violent descent in the hills outside Buxton after ascending from Congleton. On the 24th September, 1862 he took up eight passengers from a place behind Bradshawgate Chapel and he returned to the town again in June 1873 to rise from the Park Recreation Ground. By now the spectacle of a balloon ascent had ceased to attract sufficient paying patrons and this ascent appears to have been the last in the Manchester area for at least fifteen years.

The celebrations to mark the Golden Jubilee of Queen Victoria provided the incentive for the engagement of a Derby balloonist named Jackson to ascend from Ashton Market Place on the 21st June, 1887 accompanied by W.J. Andrew of the Ashton Reporter. A variable wind took them over Droylsden and Audenshaw to Reddish, then by Heaton Chapel and Heaton Norris to Didsbury where they stuck on a tree for some time before breaking free of the obstruction and finally descending at Altrincham. No planned excitements beyond the release of the balloon appear to have been provided on this occasion but when the proprietors of the old Manchester Racecourse at New Barns wished in the following year to provide an attraction they engaged a former wire walker named Baldwin from San Francisco to go up with a balloon and immediately descend by parachute. He did so successfully on two occasions and the customers must have been satisfied as, in May 1889, they were invited to pay to watch a Crewe publican named Higgins ascend before parachuting down in the company of a Miss Le Voy. Higgins repeated the performance two weeks later and, on the 16th July, he again arrived with Miss Le Voy only to find that the blustery weather conditions were not to the lady's liking. Anxious not to disappoint the crowd, Higgins replaced her with an assistant named Lennox and the balloon was released but immediately carried away on a strong wind. Higgins leapt out and safely joined the ducks in Alexandra Park Lake but of Lennox it is recorded that, after the rapidly deflating balloon had crossed Fallowfield, it crashed to earth and his body was taken into the Farmer's Arms on Burnage Lane. Higgins survived several other performances in the North before he met his death at Leeds in 1891.

Rochdale saw a number of balloon and parachute exhibitions in connection with the Infirmary Galas held from 1890 - 1895 and Baldwin was at Cale Green, Stockport on the 26th September, 1892. He ascended sitting on a sling with his parachute attached to the side of the balloon and came down at Hall Street with a dislocated arm and a thirst which he is reported to have slaked at the Blossoms Hotel. These were virtually the dying days of the balloon era with performances related far more to the realm of the circus than to any scientific pursuit. Balloons were frequently released as part of the famous Belle Vue firework displays and suspended lights were permitted to ignite the gas to explode with pyrotechnic effects. For live performances the female artist was preferred and, as well as the intrepid Miss Le Voy, Miss Alma Beaumont is reported as parachuting down to the rear of the Clegg Hall, Rochdale in 1895, Miss Maude Brookes came down relatively unharmed on the roof of the Phoenix Mill in Oldham in June 1898 for refreshments at Oldham Fire Station and two un-named ladies are said to have tried unsuccessfully to parachute on to Bolton Wanderers Football Ground in 1896.

In the early years of the twentieth century ballooning had become a popular pastime of the gentry in the South of England but the sport does not appear to have been taken up in the industrial regions of the North. Of the few ascents made in the area, only one is notable and this was made from the Botanical Gardens at Old Trafford in aid of the Lifeboat Saturday Fund. Pre-dating by many years the current fashion for despatching commemorative

Baldwin's balloon at Cale Green, Stockport on 26th September 1892. The parachute which he used for his descent can clearly be seen attached to the balloon. *:Stockport Library*

postal covers, the organisers sold large numbers of illustrated postcards which were dropped in bundles at points along the route as the balloon floated northwards crossing Haslingden and Clitheroe and coming to rest four miles from Barras in Northumberland. The aeronaut was the Frenchman Gaudron and he was accompanied by Dr Barton who is credited with the idea of the Balloon Post. Their flight was not without incident as near Settle volunteer soldiers returning from training decided to enliven the day by firing on the balloon, one shot passing through the rigging.

A few more balloons were seen on the fringes of the area, probably the most noteworthy being the *St Louis* of John Dunville, the Irish aeronaut who crossed the Irish Sea from Dublin to Macclesfield on the 15th February, 1910, but by then man's efforts were directed toward devices which would rise under their own power and travel under control from place to place and it was not for over half a century that the use of the balloon for purposes of sport and public entertainment became common again in the Manchester area.

The Pioneering Days

Whilst those men — and women — who experienced and promoted various forms of aerial travel from the nineteenth century achieved their successes by the use of various forms of aerostatic globe beneath which they floated free but at the mercy of the prevailing winds, the search for a device which would rise under its own power and proceed to a pre-determined point under the control of a pilot had continued, not only in Europe but, particularly in the latter half of the century, in America and to some extent in Australia and South Africa.

In England, the Yorkshire baronet Sir George Cayley, first applied his scientific knowledge to the use of the rotating wing which had been known centuries before as a flying plaything, and went on to form theories of aerodynamics and impulsion which led to his being described as early as 1846 as the father of aerial navigation, an assessment which, far from being dispelled, has found hardening support over the succeeding years. By 1849 Cayley was well on the way to achieving directed and sustained flight and in 1853 he projected his coachman across a valley aboard a glider sufficiently airworthy to be copied and flown with a fair degree of control more than a century later.

By 1825 the steam engine had been successfully applied to both ships and rail-guided vehicles and inventors set about considering its application to aerial transport. Two Somerset lace makers, William Samuel Henson and John Stringfellow, had produced designs for steam powered aircraft and Stringfellow's model was demonstrated, suspended from an overhead wire, at Chard and at the Cremorne Gardens in London. Inability to produce an engine light enough to be sustained by the flying machines which had been devised led away from powered flight to wide experiment to establish means of achieving aerodynamic lift and the kite and the man-carrying glider were studied with increasing vigour and success. The German Otto Lilienthal and the Scot Percy Pilcher produced practical gliders, Lawrence Hargrave in Australia and the American Samuel Franklin Cody in Britain moved toward gliding flight via the realm of man-lifting kites and in the United States two cycle manufacturers, Orville and Wilbur Wright set about developing a controllable glider which would ultimately carry its own propulsive power.

On the 17th December, 1903 Orville Wright became the first man to make a powered, sustained and controlled flight and a new era had begun.

Not everyone yet had faith in aviation, nor could the enormous development of the fixed-wing aeroplane be envisaged and as late as the 11th September, 1908 the Manchester Guardian in its leader column was to state 'We cannot understand to what practical use a flying machine that is heavier than air can be put.'

In theoretical science Manchester made a contribution in this period which was not to have its full impact until many years later. The University's first Professor of Engineering was Osborne Reynolds whose research in a converted stable in Quay Street was principally directed toward the study of turbulence in fluids. From this research he evolved the formula known as Reynolds Number. Although the science of aviation was far from the Professor's thoughts his discovery is today of vast significance to aeronautical engineers the world over.

Reynolds was succeeded at the University by Horace Lamb whose work created a foundation for modern analytical methods in fluid motion.

One of the few Manchester men who pursued their interest to the extent of physical involvement was Frederick D. Artingstall and as early as 1830 he sought to relate the power of the steam engine and the graceful wing strokes of the birds. When the Aeronautical Society was formed, he wrote an essay which appears in the Annual Report for 1866 and described how he had constructed a small model and suspended it from the ceiling of a room, built up a good pressure and turned on the steam. Two wings were attached directly to the piston and 'the machine jerked up and down, whirled round, rushed from side to side, and in fact performed all kinds of gymnastic movements within its limits (except flying) to the great amusement of the particular friends invited to witness the experiment.' Artingstall caught the model — no doubt with some difficulty — and was preparing for another trial when the boiler exploded and his friends disappointed him by the precipitate nature of their departure. A second experiment led to a similar result but Artingstall had switched his attention to a rotating-wing device, again using steam for propulsion, and did manage to obtain lift from the wings before they parted company with the engine. He presented his specification for a Flying Ship to the Society in 1867 and continued to study the birds, writing in 1873 of his conviction that flapping-wing flight was within the muscle-power of a man. He died in 1876, his home then being at 248 Collyhurst Road where his widow was said to have been left totally unprovided for, largely as a result of his scientific pursuits.

Artingstall may well have been one of the first Manchester men to become known beyond the immediate vicinity of the city for his experiments in aviation but it does appear that another experimenter aspired to local fame at a much earlier date. In 1693 Moses Cocker, a Rivington farmer and a well-known Presbyterian, is said to have leaped from the roof of his barn, borne up by his faith and a set of home-made wings. His survival is said to have owed more to the texture of his manure heap than to either his faith or his wings.

The kite has been known to man for something in the region of 3,000 years, stemming from the early Chinese civilisation, and no doubt Manchester children have played with its variations over quite a long period. In 1827 Charles Pocock succeeded in travelling between Bristol and Marlborough in a carriage pulled by a kite and in July 1828 he appeared at the Liverpool Regatta and was several times drawn about the Mersey in a boat with two kites attached. Word of this event quickly reached Manchester and William Yates of Dale Street attached a pair of kites to a low pony phaeton. Accompanied by a lady and a young lad, Yates caused great excitement by travelling from Stony Knolls in Salford, down Greengate and across Blackfriars Bridge and thence via Piccadilly to Ardwick where he hauled down the kites which folded into a small package for transport.

Yates used two kites, one flying 200 feet above the other but when some years later a Denton tailor named Moorhouse sought to use a kite to lift himself from the Belle Vue Gardens he put his faith — up to a point — in a single surface 20 yards high. On the 4th September, 1842 he exhibited the kite in the gardens and then sent it aloft with three 56 lb weights attached but, although this experiment was apparently successful, he deemed it prudent not to ascend himself.

The Manchester area in the second half of the nineteenth century appears to have produced little of note in furtherance of aviation, although an entry which appeared in the Aeronautical Journal in 1897 reported the registration of a flying machine patent in the name of the distinguished American experimenter, Octave Chanute of 70 Market Street, Manchester. Chanute travelled extensively in Europe and he may have visited the city but it seems likely that the address was merely that of a patent agent. Possibly the most significant date of the last decades of the century, at least so far as aviation was concerned, was the 26th April 1877 when the wife of a Patricroft doctor gave birth to her fourth son and gave him the name Alliott Verdon Roe, the 'Verdon' coming from Mrs Roe's maiden surname. It was not until 46 years later that the London Gazette was to announce that a hyphen had been added to the name thenceforward to be used by Sir Alliott Verdon-Roe.

The doctor's house still stands on Liverpool Road and is well maintained by the City of Salford under the name 'A.V. Roe House' given to it by the former Borough of Eccles.

One of Roe's brothers followed his father into the medical profession and it was hoped that the young Alliott would also do so. At the age of eight he commenced school at Brooklands in Surrey where he was later to become well known but as his schooling progressed his preferences were more for sporting activities than for scholastic achievement and cycling and athletics occupied much of his time. In 1892 he left St Pauls school and spent a year in Canada before returning to take an apprenticeship at the Lancashire and Yorkshire Railway works at Horwich, where for five years he worked hard both in developing his knowledge of engineering and, in his spare time, in developing his prowess as a racing cyclist, winning many prizes at meetings throughout the country. A career in the Royal Navy attracted Roe and, after a spell of work at Portsmouth Dockyard, he studied at King's College, London but was not successful in passing the entrance examination and became an Engineer Officer on merchant ships sailing to African ports.

Whilst sailing in tropical waters, Roe became fascinated by the albatrosses which followed his ship and, watching the graceful gliding flight of these birds, he became increasingly interested in producing an apparatus by which a man might emulate them. His first model gliders evoked considerable mirth amongst the crew but he persevered and succeeded in making a number of aircraft which flew satisfactorily. By 1902 Alliot Verdon Roe had already made successful models with various configurations of wing and tailplane and he filled his spare time exploring the possibilities of mechanical flying, leaving the sea and joining the motor car manufacturing company of Brothers-Crockett Ltd as a draughtsman to gain experience which was to be invaluable in later years. In 1903 he read of the work of two American cycle manufacturers named Wright who had produced a flying machine and wrote to them of his own experiments, receiving an encouraging reply. Two years later, appalled by the massive apathy of his countrymen on the subject of powered flight, Roe wrote to the Editor of The Times drawing attention to the Wright Brothers' achievements and expressing his view that a successful motor-driven aircraft could be flying over England by the summer of 1906. With heavy finality, the Engineering Editor added a footnote giving it as his opinion that all attempts at 'artificial aviation' were foredoomed to failure!

To use the idiom of a later period, Roe could not have agreed less and in March 1906 he decided to devote the whole of his time to aviation. It is possibly a measure more of his intense conviction than his undoubted engineering ability that he immediately obtained the position of Secretary of the Aero Club, as that body was then far more heavily committed to the balloon than to the aeroplane

A.V. Roe with the model which won the £75 major prize at the Daily Mail competition held March 1907.

but he was to hold the position for only a short time before travelling to America to assist in the design and construction of a large steam-driven helicopter for a man named Davidson. The helicopter was constructed but Roe had little faith in its ability to achieve controlled flight and after returning to England to conclude patent arrangements, he and Davidson parted company.

Although A.V. Roe did not follow the pattern of a number of other pioneers in producing a man-carrying glider, preferring, possibly for economic reasons, to pursue his experiments with elastic driven models, he nevertheless devoted considerable thought to the design of a practical full size machine and in July 1906 patented a form of control which would enable both lateral and fore-and-aft movement to be directed by a single unit. He seized the opportunity of entering for a contest organised by the Daily Mail and built three models which he flew at the Agricultural Hall in March 1907 to capture the premier honours from over 200 contestants. With the £75 which he received and encouraged in the belief that his designs were in the right direction, he set about constructing his first full size aircraft, based on the most successful of his models.

Roe obtained permission to build a shed at Brooklands where the motor-racing authorities had offered a prize of £2,500 to the first aviator to fly round the track before the end of 1907 and, after commencing construction at the Putney stable of his brother, Dr S. Verdon Roe, he moved his machine to the track where he fitted a 9 hp J.A.P. engine. With this he made a number of towed 'hops' but there was insufficient power to enable the aircraft to rise unaided. Shortly afterwards a 24 hp Antoinette engine arrived from France and Roe set about installing this in his biplane. The time left for the prize offered for a circuit of the Brooklands track had expired and an unhelpful management hindered progress, but in his book *The World of Wings and Things* A.V. Roe records that on the 8th June, 1908 his aircraft left the ground for the first time entirely under its own power.

Forced to leave Brooklands, Roe returned to his brother's stable and from there moved to a couple of railway arches at Lea Marshes where he flew the first of a number of small triplane designs using the 9 hp J.A.P. engine. That original triplane, which can now be seen in the National Aeronautical Collection at the Science Museum, was taken to Blackpool in October 1909 and, despite its lack of power, made a number of short flights. After returning from Blackpool, Roe moved to accommodation at Wembley Park where he decided to employ the assistance of another draughtsman on a new and more powerful triplane and Reginald J. Parrott commenced an association which was to extend well into the history of the company which took A.V. Roe's name.

For some months in 1909 Humphrey Verdon Roe had taken a keen interest in his brother's activities and, when it became apparent that outside investment in these activities was unlikely to be forthcoming, he set about forming a company with finance drawn from his own Manchester holdings. A former Captain in the Manchester Regiment and a veteran of Ladysmith, H.V. Roe had built up the business of Everard & Co, webbing manufacturers with premises at Brownsfield Mill on Great Ancoats Street, Manchester and when A.V. Roe & Co was registered on the 1st January, 1910 the basement of the mill was turned into a workshop. At the same time, the appointment of a new management at Brooklands had enabled a group of pioneer aviators to assemble there and the new company took a shed alongside them. Construction work was to be based at Manchester with the designing and testing of aircraft being carried out at Brooklands where, in June 1910, a young mechanic, Howard Pixton, joined Roe, working without pay for the privilege of having flying lessons — lessons which were to lead to him gaining Aero Club Certificate No. 50 on 24th January 1911 and subsequently winning the 1913 Schneider Trophy race.

The second Blackpool Flying Meeting at the end of July attracted a large collection of aeroplanes and a strong contingent of French pilots came to compete for prizes. Two triplanes were taken from Brooklands and loaded on a train but tragedy struck on the way and at Coppull, near Wigan, sparks from the engine ignited the tarpaulin covering their truck and the resultant fire reduced both machines to ashes. Pixton who was travelling with them was left with the unenviable task of breaking the news to A.V. Roe and he subsequently wrote of the shock with which the news was received. However, Roe was not a stranger to set-back and, by working day and night, he and his small staff produced a new machine from a miscellany of parts left over from the construction of the destroyed aircraft. Erection was completed at Blackpool, a Green engine was installed and by August Bank Holiday the Lancashire crowd was able to cheer their own man into the air. Roe himself had no illusions as to the safety and efficiency of his hastily completed machine and he survived three crashes during the meeting but a consolation prize of £75 encouraged him to continue his efforts. To celebrate, he got married before leaving Blackpool and immediately afterwards joined Claude Grahame-White for a visit to the United States, where he flew both his Blackpool aircraft and another triplane which had been despatched

The first Roe II Triplane "Mercury" at the Olympia Aero Show 1910. This was the first machine built by A.V. Roe & Co. at Brownsfield Mill.

:Flight International

to the Harvard Aeronautical Society.

For the rest of the year and throughout 1911 both the flying training and manufacturing sides of A.V. Roe & Co continued to develop. A number of trainees who were to become famous as pilots made their initial flights at the Avro flying school at Brooklands and aircraft, engines and spare parts were advertised from 'The Aviator's Storehouse' at Brownsfield Mill. H.V. Roe led the Manchester effort and handled the business affairs generally, the management of the Everard webbing concern being left in the capable hands of John Lord who was later to take a major part in Avro affairs before leaving with A.V. Roe to form Saunders-Roe Ltd and to hold various offices with the Society of British Aircraft Constructors.

As well as the triplanes and biplanes from Roe's own drawing-board, aircraft to other designs were built at Brownsfield Mill from time to time. One of the earliest was a Farman-type biplane for a Bolton man named Edwards and a Curtiss-type aircraft was built as a float-plane for Captain Wakefield who flew it from Windermere as the Lakes Water Bird. It was later joined there by a Type E biplane which took the name Lakes Sea Bird and survived as a trainer until 1915. A number of parts were supplied for the famous Irish aviatrix Lilian Bland and a monoplane was constructed for a Peruvian naval officer, Lieutenant Burga. This was a single-seater, as was the Company's only other early monoplane, the Type F built early in 1912 to try out Roe's theory that the occupants of an aeroplane need not be exposed to every whim of the elements. The Type F was the first aircraft in the world to fly with a totally enclosed cabin and Lieutenant Wilfred Parke reached 1,000 ft in it on the power of a 35 hp Viale engine. Although the Type F saw only limited use, A.V. Roe continued the cabin theme with the Type G biplane which was the Company's entry for the Military Aeroplane Competition in August 1912 winning a £100 award but failing to capture a major prize.

Before the completion of the enclosed Type G, A.V. Roe and Co had already won an order from the War Office for three Gnome-engined biplanes for use at the Central Flying School (C.F.S.). The aircraft were similar in construction to the earlier Type E and Roe felt that with the fitting of the 50 hp Gnome engine he had achieved his best design to date, justifying the institution of a new system of type numbering. The Avro 500 was the first of a design series which was to last for over 50 years and aircraft of the type were to remain in Service use well into the Great War.

For some time H.V. Roe had been concerned that Manchester people should have the opportunity of seeing Manchester built aeroplanes in the air and as early as March 1910 he had advertised in the Manchester Guardian for a suitable flying ground. In June 1912, when the third C.F.S. aircraft was completed, it was taken to Eccles Cricket Ground not far from the birthplace of its designer and tested there by Lieutenant Parke who also flew the machine at Trafford Park.

By the beginning of 1913, the former ship's engineer who had watched the albatrosses, had seen the firm which bore his name become established as a viable concern. The introduction of additional finance from the Groves family of Manchester brewers enabled the transfer of the works to new premises in Clifton Street, Miles Platting where the staff included a brilliant young draughtsman named Roy Chadwick who was to remain with the company until his death in 1947 and to achieve undying fame as the designer of the Lancaster and many other Avro aircraft. Working with Chadwick under Reginald Parrott was Harry Broadsmith who was later to represent the company for many years in Australia.

Several designs of varying significance came from the drawing boards at Clifton Street but none was to achieve greater fame than one which stemmed from a pencilled sketch by Alliott Verdon Roe in early 1913. Based upon the successful C.F.S. machines, the Avro 504 was to become the backbone of all Service flying training and to remain in production until over 8,000 aircraft of the type had been built. Initially powered by an 80 hp Gnome engine, the Avro 504 progressed to many variants and has been well described both in *Avro Aircraft Since 1908* by A.J. Jackson and *Avro — An Aircraft Album* by Eric Harlin and George Jenks.

Whilst Roe had been developing the theme which was to lead through the years to the modern aircraft produced at the Manchester factories of the successors to his Company, other aspects of aviation were also impinging upon Manchester's citizens as they read of growing activity abroad, in other parts of the United Kingdom, and eventually in their own area.

In 1906 Santos-Dumont made the first officially recognised flight in Europe; in 1908 Cody flew at Farnborough and in October 1909 the first authorised aviation meeting in the United Kingdom was organised at Blackpool by a Committee which included representatives of the recently formed — and short lived — Lancashire Aero Club. Amongst the Club's representatives the names of Mr Bowker of Radcliffe and Mr Hollingdrake of Stockport appeared and for the duration of the meeting considerable interest was shown in Manchester. The proprietors of the Manchester Guardian donated a trophy and arranged that, when flying was likely to take place, a flag would fly from the paper's offices in Cross Street to enable those who wished to take a train to Squires Gate. Interest was also shown in the meeting at Doncaster which took place at the same time and on the 25th October, 1909 Theodore Gilsdorf, general manager of a Manchester bakery firm, was reported to have paid £1,000 for the parasol monoplane produced by the French propeller maker Chauvière. The aircraft had failed to leave the ground at Doncaster but Gilsdorf who lived at 43 Birchfields Road, Rusholme was reported to have expressed the intention of installing a more powerful engine and attempting to fly in the following Spring. Nothing is known of any later appearances by the aircraft.

As Manchester grew as a commercial and industrial centre, a number of national newspapers established offices in the City. On the 17th November, 1906, before any Briton had managed to sustain powered flight, the proprietor of the Daily Mail, Lord Northcliffe, offered a prize of £10,000 for the first airman to travel from a point within five miles of his London headquarters to alight within a similar distance from the Daily Mail's Manchester office. Not more than two stops were to be made en route and the prize was equally available for a flight in the opposite direction. To add attraction to potential entrants, the Adams Manufacturing Company offered an extra £2,000 if the successful competitor flew an all-British machine and the 'Autocar' offered £400 to the maker of the winner's engine, provided that it was of British manufacture.

For the next three years the limited range of available aircraft precluded any serious thoughts of taking the prize but by 1909 a number of aspirants had signified their intention of making the attempt. George Davidson from Banchory, who at one time employed A.V. Roe as a draughtsman, had plans for a 100 seater Air Car and as an interim measure set about building a reduced version with three sets of biplane wings and two rotary lift devices at Taplow in Buckinghamshire. He submitted an entry for the London-Manchester prize late in 1909 but lack of funds prevented the completion of his machine. Eugene Gratze had built a single seat monoplane with variable incidence wings which flew in Essex in 1909 and named it the *Gratze Daisy* and he registered this entry as did Edgar Wilson with a biplane based on the Wright designs. Neither of these English entrants reached the stage of setting out for Manchester and the Frenchman Dubonnet who hoped to make an attempt delayed his preparations and found that the prize was won before he arrived on the scene.

The first serious attempt to take the £10,000 was made by the man who is generally accepted to have made the first sustained powered flight in Britain, Samuel Franklin Cody and he spent two days motoring along his proposed route and arranging for a series of smoke signals to mark his course. His original intention was to fly the route from North to South but damage to his aircraft led him to return to Aldershot for repairs and he eventually set off from London late in 1909. A seized engine led to a precipitate landing and he went back to Aldershot.

Two names are generally associated with what has become known as the London-Manchester Race, those of the Frenchman Louis Paulhan and the Englishman Claude Grahame-White. Paulhan visited Manchester in October 1909 and inspected potential landing fields at Urmston and Chorlton but postponed his attempt in deference to the approaching winter weather. Grahame-White, after arriving back from France with a new Farman biplane in the second week in April 1910, then made his own inspection of likely grounds. He paced out a field beside Burnage Station in Fog Lane and decided that this was to be his landing point.

On the morning of the 23rd April Grahame-White took off from Park Royal and followed the London and North Western Railway to a pre-arranged landing ground near Rugby where he refuelled and then flew on to land in a field near Lichfield. Here the Farman was overturned by a gust of wind and so badly damaged that it was necessary to return to Wormwood Scrubbs for repairs. Shortly afterwards Grahame-White received the news that Paulhan had reached London with a similar Farman biplane and the quest for the prize became, in effect, a cross country race between the two pilots.

On the evening of the 27th April Paulhan took off from Hendon and within an hour Grahame-White was airborne from Wormwood Scrubbs in pursuit. With the advantage of a greater period of remaining daylight, the Frenchman pressed on for 117 miles before landing at Lichfield. Grahame-White covered 60 miles and came down at Roade as the light faded. At 2.45 am the following morning he took off by the light of car headlamps in a desperate attempt to overtake his rival but increasing winds forced him to land again at Polesworth where he remained whilst Paulhan flew on, following the railway past Crewe, Sandbach, Wilmslow and Gatley to circle over Ladybarn and land at 5.32 am in the field beside Burnage Station which his rival had selected for a triumphant arrival.

Great crowds had waited through the night to greet the contestants and when Paulhan appeared he was surrounded by cheering spectators and carried to the Station to join his wife on a special train into the city. One of his first acts was to send a telegram to Grahame-White offering his good wishes and expressing the hope that he would also reach Manchester but, although Grahame-White wired back his congratulations and made ready to complete the flight, he was defeated

Paulhan's Farman biplane at Didsbury being dismantled for its return to London. Pending dispatch it was stored at the Renold Chain Works on Burnage Lane.
:Renold Limited

by the weather. When, almost twelve hours after Paulhan had reached Manchester, the wind and rain abated, he took off again but finally landed near Lichfield with the fabric flapping from his aircraft and, as a further storm broke over the field, he gave orders for the machine to be dismantled and returned by rail to London.

Paulhan slept until noon at the Queens Hotel before lunching with the Lord Mayor at Manchester Town Hall. He returned to London the same evening and his biplane was dismantled and stored temporarily at the Renold Works, Burnage before being taken south by rail. Forty years later he celebrated the anniversary of his flight by flying over the same route as a passenger in a Meteor T7 and on the 22nd October, 1962 he flew from Paris in a Caravelle of Air France to attend the opening of the new passenger terminal at Manchester Airport. When the field where he had made his historic landing gave way to a Council housing estate the name of Paulhan Road was used to mark his achievement and in 1976 a plaque was placed on one of the houses to keep the memory alive.

Grahame-White's valiant efforts were not forgotten. When the two men returned to London and were feted at a banquet at the Savoy, he received a silver rose bowl as a consolation and then accompanied Paulhan to Charing Cross Station for his return to France.

Whilst the first entrants for the £10,000 London-Manchester prize had been considering routes and possible landing grounds before diverting their attention to the Blackpool and Doncaster meetings, a Manchester motor engineer was working on a monoplane which he hoped to fly at Blackpool. Charles Fletcher of Heald Grove, Rusholme had a works on Stockport Road where his company, the Empress Engineering company, built motor cars and produced the country's first motor ambulance which was supplied to Salford Corporation. In association with Norman Crossland of Prestwich and William Arnold of Upper Brook Street, founder of the motor business bearing his name which is still well known in Manchester, Fletcher took as apprentices his son and also a young lad named John Alcock who first distinguished himself by backing the firm's heavy lorry through the wall of the workshop.

The monoplane was a single seat tractor type with features drawn from both Antoinette and Bleriot and was powered by a five cylinder rotary engine which Fletcher built himself. On the 20th October, 1909 it was reported to have flown 30 yards in Heaton Park. Trials in the park continued and on one run the aircraft is reported to have been damaged in collision with the bandstand. It was mentioned in 'The Aero' as having made 'several long jumps of 75-100 yards' during January 1910 but on the 17th of that month a gust of wind caught it as Fletcher was about to take his seat and violent collision with a clump of trees caused considerable damage. An Empress monoplane was reported as being exhibited at the White City in March and it appears that this was a second aircraft, possibly with the engine and other components which survived the wreck.

Fletcher turned his attention to the construction of a biplane but continued to develop his engines, one of which was tried in an Avro Triplane. Both five- and seven-cylinder engines were produced. In August 1910 the Empress biplane made several long hops at the racecourse at Castle Irwell in Salford piloted by Fletcher and possibly also by John Alcock but after alighting in the river on at least one occasion and on another demolishing the course manager's cucumber frames it ceased to be acceptable to the management and Fletcher, accompanied by Alcock, took his machine to Brooklands in Surrey where many of the country's foremost experimenters foregathered. Alcock became friendly with the Frenchman, Ducrocq, who employed him as a mechanic and taught him to fly, his Royal Aero Club Certificate being numbered 368.

The biplane passed from Fletcher's hands and was subsequently flown at Brooklands by Macfie and Valentine. The Empress Engineering Company ceased to exist shortly afterwards following a disastrous court action arising from the alleged infringement of Gnome engine patents and Alcock pursued his career at Brooklands.

Although Fletcher was probably the first to make serious and partially successful attempts to fly in the Manchester area, the number of local men who had read of the successful feats of aviation elsewhere and aspired to emulate them was growing. The Hon. Maurice Egerton, later to become Lord Egerton of Tatton, bought the fourth Short-built Wright biplane, S.35 and flew at Eastchurch, receiving aviator's certificate No. 11 in 1910, and later becoming the first to use wireless equipment on an aeroplane. Others less endowed financially sought to band together to further their interest in aviation and on the 25th August, 1909 a preliminary meeting at the Douglas Hotel on Corporation Street under the chairmanship of Charles Stevenson made arrangements for a grand inaugural meeting of 300 people at the Midland Hotel on the 9th September when Sir William Bailey took the chair and mentioned plans for an aerodrome at Trafford Park. Bleriot, Cody, Grahame-White and A.V. Roe were invited to become Honorary Members and the members of the Committee included Cedric Lee, G. Tilghman Richards and H.V. Roe who was also to become Chairman of the Club's Model Section.

By the 2nd November, 1909 the Manchester Aero Club had inaugurated a series of weekly lectures by Mr L.B. Desbelds at the Manchester School of Technology and on the 4th March, 1910 the Club organised an exhibition of aeroplanes and models opened by the Lord Mayor at the White City. The following full-size aircraft are listed in the programme as being on show: —

Bleriot Cross Channel Type Monoplane
Kindly lent by Mr Leo Swain, Messrs Bleriot's Manchester agent.
Fletcher Monoplane
Kindly lent by Mr Fletcher, Empress Motor Works, Manchester.
Avroplane
Kindly lent by Messrs H.V. Roe & Co(!), Brownsfield Mill, Manchester.
Staniar Glider
Kindly lent by Mr John Staniar, Lansdowne Road, Crumpsall, Manchester.
Glydoplane
Kindly lent by Messrs Threlfall and Leigh, 30 Withy Grove, Manchester.

Threlfall, Leigh and Staniar were active members of the Club but do not appear to have been successful in coaxing their machines into the air as no further record of the machines has survived. The Roe Triplane was christened *Mercury* at the exhibition, Miss Behrens, daughter of the Lord Mayor, performing the ceremony.

The Bleriot XI which was exhibited was described as that which had previously been displayed in Messrs Cockshoots' showroom at the time of the Manchester Motor Show at the City Hall when a Fletcher monoplane was suspended from the roof of the hall. Mr John Norris who was for many years Managing Director of Joseph Cockshoot & Co Ltd asserted in later years that this aircraft was borrowed from London and returned there after the Motor Show.

However, it may have been the machine which was purchased by Leo Swain at approximately the same time and which, after the Show at the White City, was displayed in the window of Swain's premises on Deansgate. Leo Swain was an enthusiastic member of the Manchester Aero Club Committee and purchased the Bleriot with the intention of flying it but met with determined opposition from his wife. In April 1910 he sold it to Gerald Higginbotham of Macclesfield.

In May 1910 a Roe Triplane was displayed at the Manchester Industrial Exhibition at Rusholme, winning a gold medal and the same venue was used for a model show run by the Manchester Aero Club in the following September, shortly after the Club had formally affiliated to the Royal Aero Club. Active negotiations for an aerodrome site at Trafford Park were still continuing and by October it was announced that the ground had been inspected by officials of the Royal Aero Club and approved as an official landing place for the £10,000 Round Britain Race the following year. On the 19th December at a meeting at the Midland Hotel a new company, Manchester Aerodrome Ltd, was formed.

The aerodrome was sited on Ashburton Road and took in a large area of what had formerly been the deer park of Trafford Hall. It faced the works of Taylor Brothers & Co Ltd and had a frontage onto Mosley Road. Work to prepare the ground was pressed on during the first half of 1911 and in May plans were made for the Willows airship *City of Cardiff* to fly there. Willows, however, demanded that a hangar should be erected for his use and it was felt that available funds could be put to better use. The *City of Cardiff* would not have been the first airship to visit Manchester as a small dirigible was demonstrated in the Free Trade Hall in May 1903 and on the 11th September 1909 Spencer gave a display at the White City before flying from there to Sankey. It is believed that his airship was on exhibition at the St James's Hall in Oxford Street shortly before this.

Although Paulhan had flown into the area, no airman had yet made a journey away from Manchester and the first to do so was Henry G. Melly who had established the Liverpool School of Aviation. On the 7th July, 1911 he took off in a Bleriot accompanied by A. Dukinfield-Jones after first telephoning to A.V. Roe to arrange for white sheets to be placed on the Manchester Aerodrome to mark the landing area. Flying along the line of the Ship Canal, Melly had an uneventful trip, lunched with Roe at Trafford Park and took off again to return to his base at Waterloo. This flight did not bring any financial reward, although in 1909 Sir John Hartley had offered a prize of £1,000 for the first Liverpool-Manchester flight. Cody made an attempt to win the prize on the 29th December, 1909, getting no further than Prescot, and Compton Paterson also made an abortive attempt but the time limit set for the flight had passed when Melly made his visit to Manchester.

On the 22nd July, 1911 twentyone aircraft assembled at Brooklands for the start of the 1,010 mile Circuit of Britain, the prize being another £10,000 donated by the proprietor of the Daily Mail. The leading contestants were scheduled to reach Manchester on the 25th July and by 7.00 am the crowds began to pour into the aerodrome in scorching summer weather. Melly had been engaged to give flights with his Bleriot and he arrived from Liverpool to land before a large assembly. Flights of five minutes duration at two guineas each had been advertised but Melly had the misfortune to crash on his first flight, neither he nor his passenger, C.H. Werner, being injured. The ultimate winner of the race, Lt Conneau (using the pseudonym 'Beaumont') arrived from Carlisle after the weather had broken and he landed in rain to be briefly feted before taking off and flying over Chorlton, Styal and Macclesfield in his Bleriot on the next leg to

Bristol. Hard behind him came the Deperdussin of another Frenchman, Vedrines, who had to land near St Helens to ask his way to Manchester but the crowd were to see no more race aircraft that day. Only two others managed to complete the course, third man home being James Valentine of Mottram Hall, Prestbury who arrived at Trafford Park on the morning of the 29th July. Samuel F. Cody in a biplane of his own design arrived on the 2nd August.

For Cody the Manchester area was not a particularly auspicious place. He first visited the district with his Wild West Show and between performances demonstrated his man-lifting kites at Rawtenstall in December 1902. He also tested kites at Tyldesley and later made plans to fly to Manchester for both the London-Manchester and Hartley prizes. He was unsuccessful in each of these ventures and, although he gamely battled round to complete the Round Britain course, the race could hardly be considered a success for him. He was killed at Farnborough on the 7th August, 1913.

The Manchester Aero Club attempted to form a flying school at Trafford Park in February 1912 but the aerodrome was not much used and in December of that year the Manchester Aerodrome Company relinquished the tenancy of the field. On the 14th May, 1913 George Lee Temple brought his Anzani-powered Caudron biplane to give a number of demonstrations and on the 20th June, 1914 the aerodrome was again used for a major air race, this time from London to Manchester and return. Eight aircraft started at Hendon and four dropped out of the race before reaching Manchester. First to arrive was Louis Strange and he was greeted by the Lord Mayor, Alderman McCabe, who climbed onto his Bleriot to shake his hand and so damaged the aircraft that it was unable to take part in the second part of the race. The three remaining pilots landed and took off again

John Alcock's Maurice Farman Longhorn at Trafford Park on 20th June 1914. This aeroplane had an early Sunbeam engine the radiators for which can be seen round the nacelle. *:E.A. Harlin*

This monoplane built in 1911 by Hill and Co. of Bury appears to have been heavily influenced by the Bleriot XI design. *:P. Lewis*

safely, the winner being W.L. Brock who, like the runner-up, R.H. Carr, flew a Morane-Saulnier monoplane. The third finisher was John Alcock in a Sunbeam-engined Farman. Brock's machine was later taken to the United States and it is believed that surviving parts are being used in a reconstruction of the aircraft. Undaunted by his contretemps with the Bleriot, Alderman McCabe later went up with P. Verrier to become the first Lord Mayor to fly over the environs of his City.

Whilst the Manchester Aero Club attracted the support of most of the district's major flying enthusiasts, the towns around Manchester produced a number of groups of keen aircraft modellers and Oldham, Smithills, Higher Broughton and Macclesfield each had its Model Aero Club. Building was not confined to models and at Bury Hill & Co produced a monoplane bearing a close resemblance to a Bleriot XI. South of the city Charles Pickering and Norman Willoughby of Knutsford, both still in their teens, constructed a monoplane glider on the lines of the Barber Valkyrie with a 26 foot span and C.C. Horner of Macclesfield evolved a graceful model sesquiplane. Also at Macclesfield, Melville Stone commenced construction of a biplane and Gerald Higginbotham of Ivy Lane experimented with the Bleriot which he had bought from Leo Swain.

Higginbotham was the first man to drive a motor car in Macclesfield and became a director of the Belsize Motor Co. He was a keen experimenter and had, amongst other things, produced X-ray apparatus and a powerful searchlight. Amongst his motoring achievements was the driving of the first car to reach the summit of the Great Orme at Llandudno and he only narrowly failed to drive up the 108 steps in the centre of Macclesfield. Higginbotham towed his Bleriot from his home to a field beside Congleton Road and succeeded in becoming airborne but found that he could not get over the surrounding hedge. He took the machine to Freshfield and learnt to fly, obtaining Royal Aero Club Certificate No. 96. His later flying was mainly in the Liverpool area but he maintained a large workshop at his home and there he built a variant of the Bleriot with a steel fuselage. Reports of an aircraft which is said to have become airborne at Chelford in June

1910 but to have crashed on landing may have related to this machine.

Whilst Higginbotham was flying at Freshfield, George Tilghman Richards, who had also had connections with the Belsize Motor Co before setting up his own business at Broadheath, joined with another member of the Manchester Aero Club, Cedric Lee, to study the construction of aircraft with annular or circular wings based on a biplane design evolved by a Lancaster man named Kitchen. It does not appear that the machines ever flew in the Manchester area but a monoplane built to Lee and Richards designs was flown successfully at Shoreham. A replica of the first annular biplane was made for use in the film 'Those Magnificent Men in Their Flying Machines' and still survives in the collection of the Newark Air Museum.

As well as the more conventional designs, the period before the Great War saw several less orthodox inventions. In 1912 William Booth of Eccles was offering the patent rights of a self-stable aeroplane which, had it been successfully developed, might have solved the problems of many later designers. In those days there was no Trades Descriptions Act and Booth advertised 'ailerons and wing-warping superseded, side slip conquered, automatic recovery from angle of banking, no dihedral angle necessary, models answer any test.' Booth claimed to have discovered the secret of the bird's natural lateral stability and produced models of an odd machine with twin propellers but did not achieve a full size aircraft. Also in Eccles the firm of C. Binks Ltd of Church Road advertised their willingness to undertake the design and manufacture of aeroplanes but do not appear to have produced any aircraft. Joseph Clarkson of Prestwich Park North produced a model of his Clarkson Aerocar which was to have had four direct lifting aerial paddle wheels to solve the problem of vertical flight but, although the paddle wheel was demonstrated at the Manchester Municipal School of Technology and the model was demonstrated at Olympia in 1911, no full size machine was built.

Clarkson was not alone in placing his faith in rotary devices and, as early as September 1909, S. Lipman of Longsight had produced a model with '6 Helixes and 2 propellers' and claimed that a vehicle to carry two passengers could be built for £500. Also interested in vertical ascent and descent was Thomas Johnson, a Radcliffe grocer who built a bird-like model and proposed an aeroplane which would rise without any preliminary run. In the event of a mishap, a parachute arrangement would come into operation automatically to bring the aircraft down safely. Man-powered flight also had its adherents and in 1910 a cycleaeroplane said to have been built by a German workman was advertised for sale by a Wigan man.

Not all of those who were far-sighted enough to recognise the aeroplane's potential were concerned to construct their own machines. A.V. Roe and Co Ltd did so but also gave notice that their 'Aviator's Storehouse' at Brownsfield Mill could supply virtually everything required by the aspiring constructor. W.H. Cowley & Co Ltd of Daubhill, Bolton made engines and Jones Brothers Ltd of York Street, Manchester advertised in 'The Aero' cotton fabrics for aeroplanes. Propellers and spars could be obtained from Wild Bros of Daisy Bank Works, Salford and five months after their works in Burnage had housed Paulhan's Farman, Hans Renold Ltd won a Diploma of Honour for their products in the aeronautics section at the Brussels Exhibition.

Whilst local men developed their own talents, others visited the district to give flying displays and as early as July 1910 George Barnes brought his Humber monoplane to provide an additional attraction at the Muratti Gold Cup cycle

Gustav Hamel's Bleriot at Garner's Lane, Adswood 6th September 1913.
:Lancashire Aero Club

meeting at Fallowfield and gave a brief demonstration heading away from the track towards Levenshulme. Two years later Claude Grahame-White brought his Grahame-White Baby to the same ground and made two short flights on one of which he landed in Birchfields Park.

Gustav Hamel who was drowned in the English Channel in July 1914 also visited the area on a number of occasions. He flew his Bleriot from Macclesfield to Buxton on the 27th August, 1912 and gave a demonstration at Daisy Nook, near Ashton, on the 12th March, 1913, following this with a display of aerobatics at Adswood on the 6th September. Robert Slack, also flying a Bleriot, made a tour of Britain in July 1912 calling at Didsbury before travelling on to Southport.

On the 1st January, 1914 an exhibition which included the Roe I triplane and the fuselage of the Avro 508 was opened at Belle Vue and, as well as Fred Raynham in an early Avro 504, B.C. Hucks gave demonstrations of aerobatics from the Sports Ground in the gardens flying his 80 hp Bleriot *Tornado*. After a number of displays spread over three days, Hucks set out to fly to Leeds on the 4th January. He ran into a snowstorm and landed on Lobden Moor. The aircraft was dismantled and taken to a barn at Nick Road Farm where it remained for several days whilst Hucks hoped for an improvement in the weather to enable him to continue his journey but eventually he gave up and the aircraft was towed back to Manchester behind a milk cart.

The First World War

When Manchester awoke to the realisation that Britain was at war, there were probably very few people who envisaged the length of the conflict or the extent to which the role of the aeroplane would be developed in the years which were to follow. In 1908 the Manchester Guardian had foreseen very little future for the aeroplane and had gone on to express the view that it would be of no conceivable use in war. Despite the efforts of the dedicated men who formed the Royal Flying Corps and the Royal Naval Air Service, both the General Staff and the general public were to need some time to accept the fallacy of that view.

The limited range of the aircraft of the day militated against the establishment of operational air bases in the Manchester area but it was a unit formed in Manchester which distinguished itself in one of the earliest raids against the enemy. Germany had already developed the rigid airship as a weapon of war and there was considerable apprehension about the potential effect of aerial bombardment both against military installations and against the virtually undefended cities of Britain. With little prospect at that time of intercepting and shooting down the gas-filled leviathans, the Royal Naval Air Service turned its attention to attacking the bases where they were kept and in great secrecy a squadron of four Avro 504s, powered with the 80 hp Gnome engine, was formed at Manchester with two pilots and eleven air mechanics. The aircraft numbered 179, 873, 874 and 875, were taken new from the factory, crated and carried by train to Southampton where they were placed on board the *Manchester Importer* and taken to Le Havre. From Le Havre a special train took the aircraft to Belfort in Eastern France where they arrived on the night of the 13th November, 1914.

The aircraft were erected the following day and the improvised bomb release gear designed by A.V. Roe was tested in preparation for an attack on the airship sheds at Friedrichshafen on the northern shores of Lake Constance, close to the borders of Switzerland. Although the aircraft were ready, the late arrival of additional pilots delayed the start of the operation and deterioration of the weather conditions precluded any flying for some days.

On the morning of the 21st November the aircraft were each fitted with 4 x 20 lb. high explosive bombs and tuned up for the 125 miles trip to their target. Three aircraft took off successfully but No. 179 failed to become airborne and remained at Belfort with a damaged tail-skid. Led by Squadron Commander E.F. Briggs, the three aircraft crossed the Black Forest mountains and attacked the airship base, their small bombs setting the gasworks on fire and heavily damaging the Zeppelin works. Heavy small arms fire was encountered and Briggs was shot down and captured but Flight Commander J.T. Babington and Flight Lieutenant S.V. Sippe returned safely. All three pilots continued to serve with the Royal Naval Air Service, Briggs after escaping from enemy hands subsequently commanding the experimental station at Eastchurch. Babington went on to become a Lieutenant Colonel in the Royal Air Force flying Handley Page 0/400 bombers and Sippe, who had learnt to fly with A.V. Roe in 1911, ended the War in command of the Air Acceptance Park at Brough.

The Avro 504 was to be used principally in the training role but the aircraft had its brief period of operational glory and a number were fitted with guns

Flt. Lt. Sippe's Avro 504 (874) about to take off for Friedrichshafen. Flt. Lt. Cannon's machine (179) in the background broke its tailskid and was unable to make the flight.

to contribute to the defence of the country against the Zeppelin raids. It was, however, again with 20 lb bombs that a 504 was to achieve a further conspicuous success when, on the 17th May, 1915, Flight Commander A.W. Bigsworth of the Royal Naval Air Service in No. 1009 struggled to 10,000 feet over Ostend and succeeded in dropping his four bombs on the LZ 39 so damaging the airship that it was forced down at Evère and wrecked.

Production of 504s continued at Clifton Street but it soon became clear that further space would be needed. Providentially, a large extension to the Newton Heath works of Mather and Platt Ltd had been built but not occupied and A.V. Roe & Co Ltd took over the premises. A triangular piece of land close by was purchased and, after a great deal of work to fill in and level the ground, a large factory was built but production in the new Newton Heath Works was not established until after the end of the War. Design work fell more and more under the control of Roy Chadwick, aided by F.W. Vernon, J. W. (Jock) Ratcliffe and Harold Rogerson.

To anticipate an expected demand for large seaplanes Roe persuaded the Company's board to purchase an area of land at Hamble on Southampton Water, and built a works there. Although the factory was built and drawing offices established there, limitations placed on the use of building materials restricted the building of an ambitious garden village which Roe had planned and Hamble was used principally as an assembly and experimental station for aircraft manufactured at Manchester. R.J. Parrott took over as General Manager there and he was later succeeded in that office by the Rev. E.V. Roe who subsequently became Vicar of Hamble Parish Church. Amongst the personnel was a young mechanic named Roy Dobson who travelled to various aerodromes to which Avro aeroplanes were taken for flight trials. He distinguished himself on one occasion in 1916 when flying in the twin-engined Avro Pike by climbing precariously along the top decking of the fuselage from the rear cockpit to the bow gunner's position to restore stability to the aircraft which had taken off with the centre of gravity too far to the rear. The pilot of the Pike on that flight was Fred Raynham who had been a pupil at the Avro Flying School and went on to establish a fine reputation as a test pilot. The young mechanic went on to become works manager at the Manchester factory and to be knighted for his services to the aircraft industry in two wars.

As well as the Pike which was developed into the Type 529 for both long range fighting and bombing duty, large single-engined aircraft were constructed in prototype form for bombing duties and the Avro Scout Type 511 was built to compete for RFC orders. After being tested by Raynham from Southport sands, this aircraft like the others failed to attract a contract for production, the War Office preferring to leave A.V. Roe & Co to concentrate on building up its manufacture of the Avro 504.

Tight censorship prevented news of wartime flying activities reaching the newspapers in any great detail and there are few records of flying in the Manchester area at this time. An aircraft which crashed in the Peak District on the 13th December, 1914 is likely to have been in transit to one of the stations in the eastern half of the country but little information was released.

Although the company Manchester Aerodrome Limited had gone into liquidation in 1913, the field at Trafford Park continued to find limited use and various RFC aircraft, including BE2cs and FE2bs, visited. A.V. Roe & Co Ltd built a shed there and various aircraft were tested, Fred Raynham and Harry Broadsmith who had joined the Company in 1912 being there in December 1915 with the Clerget engined Avro 521 prototype which ultimately met its end in a fatal crash at the Central Flying School.

As well as the Avro shed, a building was erected at Trafford Park for the Central Aviation Co Ltd which had been formed by two Wilmslow men largely to further the aviation interests of their sons, John Oliver and Charles Brown. The Company's first aircraft was the Anzani-powered Deperdussin which W.H. Ewins had flown at Hendon and this was used for practice flying at Wilmslow before the Company moved to Trafford Park. As well as the Deperdussin, the shed housed a 2-seat Bleriot with a 60 hp Alvaston engine and Brown and Oliver who had been Avro apprentices commenced the construction of a monoplane and a biplane designed jointly by Oliver and another former Avro apprentice, Joe Bland. Neither project reached completion as both Brown and Oliver joined the Royal Flying Corps, Brown being killed shortly afterwards in France whilst flying an FE8. Oliver went on to a distinguished career in the Royal Air Force, retiring with the rank of Air Commodore in 1946 to join the Goodyear concern and as late as 1970 he was still an active pilot.

Although the Kaiser had in January 1915 authorised the commencement of airship attacks against the cities of England, it was not until a year later that the North West of England received its first visit when the Zeppelin L13 wandered over Macclesfield and Buxton on an abortive foray against Liverpool. Navigation of the airships was often wildly inaccurate and the captain of another Zeppelin reported that he had been over Manchester on the same night although, in fact, the town which he had seen was Nottingham. German reports of bombs in Manchester were not borne out by fact.

Probably Manchester's first casualties from Zeppelin attack occurred on the night of the 31st March/1st April, 1916 when 29 soldiers of the Manchester Regiment were killed by bombs which fell on the Baptist Chapel at Cleethorpes. The men were from B & K Companies of the 3rd Battalion and had been on guard at a wireless station. They were billeted in the chapel and adjacent empty shops.

The first bombs to fall in the local area were dropped on the night of 25th September, 1916 when the German Naval airship L21 crossed the Peak District and, with the captain believing that he was over the factories of Derby, released seven bombs which fell at Holcombe. To this day the marks where fragments of the bombs struck the wall of a cottage near the church can still be seen. The L21 passed on to Bolton and dropped a further 9 bombs which fell upon houses in Kirk Street, killing 13 people. The L21 made several other raids on England reaching Macclesfield on the 27th November without attacking the town. She was shot down by aircraft of the Royal Naval Air Service as she crossed the coast, falling in flames into the sea off Lowestoft.

Manchester was beyond the effective range of German bomber aircraft but the danger of airship attack led to efforts to reduce the number of lights visible from the air and on the 15th February, 1917 an air raid warning control system was established. The means of warning the public included the firing of maroons from mortars set up at strategic points. It is reported that a demonstration in Ashton Market Square produced deafening detonations 800 feet above the town.

Throughout 1917 the factories of A.V. Roe & Co continued to increase their production of aircraft. The Royal Flying Corps set up the School of Special Flying at Gosport under Lt Col Robert Smith Barry who was adamant that the Avro 504 was by far he most suitable aircraft for training purposes. The success of the Avro 504 in use at the School of Special Flying led to its increasing use in the growing flying Services as Smith Barry's Gosport system of training was adopted for all training units and production increased apace as aircraft of the type were sent to stations throughout the United Kingdom and also overseas. Shortage of operational aircraft led to the use of 504s by a number of Home Defence units until the end of the War and as production of airframes outstripped the supply of engines a universal mounting was designed to enable any available rotary engine to be fitted. The versatility of the pre-war design may be measured from the fact that throughout its long career the Avro 504 flew with no less than 20 different types of engine.

Numerous other manufacturers were pressed into service to turn out the 504 and as well as producing new aircraft the Miles Platting factory also took in for repair machines already in service with the Royal Flying Corps. Many completed aircraft were taken away for final assembly, being towed through the streets on their own wheels to be loaded for rail transport, but some were tested locally and one of the Service pilots attached for this work was an Australian, Frank Roberts, who has claimed credit for the idea of using a small field beside the factory for delivery of aircraft for modification and repair direct to the place where they were to receive attention. Roberts recalls the approach over railway yards and telephone wires and anyone familiar with the highly industrialised nature of the area will have no doubt as to the hazardous nature of this undertaking. J.C.C. Taylor, who had commenced work as an apprentice at the works in 1913 and who subsequently became Chief Engineer of Avro Transport Services and the highly regarded Secretary of the Avro 504 Club, also recalled flying from the field and wrote of flying aircraft from a cinder track beside the factory. Not all Roberts' piloting exploits were confined to aircraft and it is recorded that, with a crew consisting of Roy Dobson and Harry Broadsmith, he distinguished himself by driving a steam roller through the streets of Manchester on Armistice Night in 1918.

The Avro 533 Manchester, a promising design which was overtaken by poor engines and the cessation of hostilities. *:A.V. Roe & Co Ltd*

Many distinguished pilots first took to the air in an Avro 504 and amongst them was the man who led the nation through the dark days of World War 2. As Prince Albert, H.M. King George VI obtained his wings under the instruction of the man later to become Air Chief Marshal Sir Alex Coryton, using an Avro 504K serialled H2314.

As well as devoting his energies to the constant updating and modification of the Avro 504, Roy Chadwick found time to design the Type 530 reconnaissance aircraft with a 200 hp Hispano-Suiza engine and a small single seat fighter, the Avro Spider. Both designs were built but did not progress beyond the prototype stage. No more successful was the first Avro Manchester, a progressive development of the Type 529, its excellent flying qualities being bedevilled by the shortcomings of its Dragonfly engines — a remarkable parallel with its Second World War namesake.

The war, which some had thought in 1914 would be over in a matter of months dragged on as hundreds of thousands died in the static campaigns in France and Belgium. Manchester, as well as by now contributing its quota of men to the flying service, set out to increase its efforts to provide them with the aircraft needed for their training and for use against the enemy. To give the workers a tangible idea of the uses to which aircraft produced in Britain were put and to raise money for war charities, an exhibition was opened at the Whitworth Institute on the 26th May, 1917. Included amongst the exhibits were the Morane 'L' Parasol 3253 used by Flight Sub Lieutenant Warneford, VC to shoot down the Zeppelin LZ37, an Albatros shot down by Captain Albert Ball, VC and the observation car of a Zeppelin brought down in East Anglia. The Lord Mayor, introducing Major Godfrey Paine who opened the exhibition, spoke warmly of

Manchester's air defences and it was clear that attack from the air was still very much in the minds of the people of the City. Nor were they or their representatives disposed to suffer the raids without retaliation. On the 8th October the City Council passed a resolution calling for the immediate bombing of German towns and a reward of £1,000 was offered to the first British airman to drop a bomb on Berlin.

There is no reason to believe that the sabre-rattling of Manchester's leading citizens had anything to do with the foray made on the night of the 19th/20th October when the captain of the L41 claimed to have observed hits on the City. Once again the standard of his navigation failed to match the enthusiasm of his report — his bombs fell at Birmingham. Little more accurate but sadly more destructive in the area was the captain of the L61 who made the last raid on the North West on the 12th/13th April 1918, setting course for Liverpool and passing over New Mills and Poynton before turning away from his target as he reached Runcorn and heading northwards. Sighting the glow of blast furnaces he decided that he was over Sheffield and unloaded his full complement of bombs. The town beneath him was, in fact, Wigan and casualties were caused at Ince, in Wigan itself and at Aspull. The airship then headed for home across Radcliffe, Oldham and Royton, and although chased by a Yarmouth-based F2A flying boat, succeeded in returning to Germany.

If the impact of the airship raids was to fall with only limited violence on South East Lancashire and North East Cheshire, it at least served to spur the Home Front to renewed efforts in the factories and in the provision of funds towards the prosecution of the War. Manchester made its contribution to the Imperial Air Fleet collections when a Sopwith Camel was handed over for service

Royal Aircraft Factory SE5a D6967 which was presented to the Royal Flying Corps by the people of Horwich.

with the forces of India. The aircraft was presented by the President of the Chamber of Commerce, Mr R.B. Stoker at a ceremony at the Manchester Athletic Club's ground at Fallowfield on the 2nd March, 1918. An RE8, E254, carried the name *Marple* testifying to the savings efforts of the people of that town and an SE5a bearing the serial D6967 had the title *Horwich* painted on the fuselage.

Shortly after the presentation of Manchester's aircraft to representatives of the Indian forces, the Lord Mayor flew over the City in a locally built aircraft scattering leaflets exhorting the populace to greater efforts in support of war savings. He took off from Trafford Park and his circuit of the area was enlivened by forced landings at Heaton Chapel and Openshaw. The Lord Mayor continued to take an interest in promoting support for the flying services which merged on the 1st April to form the Royal Air Force and on the 17th June he presided at a lantern lecture at the Free Trade Hall when an anonymous Service pilot recounted his experiences. A poster advertising that meeting can be seen on display in the Royal Air Force Museum.

Other districts joined enthusiastically in promoting the war savings campaign and at Wilmslow a small exhibition was supported by Lieutenant John Oliver, who, on coming on leave from the Central Flying School at Upavon where he was an instructor, was pressed into service to give aerobatic displays at low level down the middle of Grove Street — an opportunity which he seized with alacrity! His Avro 504 flew from a small field near Wilmslow Station and, for a display of low flying which in other circumstances might have earned him a Court Martial, he was delighted to receive a letter of thanks from the Wilmslow Council.

As the war progressed through its final year there were many ways in which its effects struck home to the people of the district around Manchester — not least in the horrific numbers of young men fated never to return to their homes — but, as in the early months, so in the last the main theme of aviation activity centred upon production.

The Gorton firm, Crossley Bros Ltd commenced motor car production in 1906 after already having had several years experience of building engines and by the end of 1916 had already supplied large numbers of tenders to the Royal Flying Corps. In addition, production of Beardmore and Bentley Aero engines was undertaken. Wartime expansion of production had led to the acquisition of premises at High Lane, Heaton Chapel and when, in early 1917, the decision to build three National Aircraft Factories was taken a newly formed company, Crossley Motors Ltd, was appointed to run one of them. Further land adjacent to High Lane was taken and to give the extended lane a more pre-possessing title the City Paving Committee resolved that it should hence-forward be known as Burnham Road. This proposal stood for some months before an approach was made by Mr (later Sir) William Letts of Crossley Motors who was concerned that Burnham might be too readily understood as 'Burn 'em' and hence be inappropriate as the address of an aircraft factory! From that time onward the lane has been Crossley Road.

Contractors for the new National Aircraft Factory No 2 were H. Matthews & Sons (Builders) Ltd and work commenced in October 1917. Within nine months a 15 acre factory had been erected to standards which led Lord Weir to describe it as the finest of its class in the world and 2500 people were employed in meeting orders for de Havilland DH9 and DH10 bombers. Part of Cringle Fields park was requisitioned and, with the addition of land at its southern end, became an airfield for testing and delivery of the aircraft produced. The field was, in fact, not extensively used, although some flying did take place, and most of the aircraft

DH9 production at the No. 2 National Aircraft Factory, Heaton Chapel.
:Entwistle Thorpe & Co Ltd

made left the factory by road. The first aircraft delivered, a DH9 bearing the serial D1001, left on the 16th March, 1918 and was taken to No 2 (Northern) Aircraft Depot at Coal Aston near Sheffield after being towed to railway sidings in Broom Lane, Levenshulme.

To train workers for the National Aircraft Factory No 2 and other local factories, a Government Instructional Factory for Aircraft Cnstruction was set up at Bellé Vue and works on parts for the DH9s was undertaken in the King's Hall and a skating rink.

Considering the limited period of time between the start of building operations and the drastic curtailing of production by the Armistice on the 11th November, 1918, it was a considerable achievement by Crossley Motors Ltd to turn out something approaching 400 DH9s and, although the later production of DH10s never reached any spectacular rate, at least seven of these large twin-engined aircraft left the factory. The first two produced were both test flown from the aerodrome beside the factory and both came to grief before reaching the RAF. F351 crashed at Dunham Massey with the loss of its crew and F352, piloted by Flight Sergeants Moss and Ballard, finished in a crumpled heap on the aerodrome on the 18th February, 1919. Although both pilots were injured, they both recovered and Mr Moss subsequently became a well-respected member of the Manchester Branch of the Royal Flying Corps association after devoting much of his time during the Second World War to work for the Air Training Corps.

Whilst the Avro works and the National Aircraft Factory No 2 bent their efforts to the production of aircraft for the British flying services, steps were being taken to provide another large production facility in the area, this time to produce aircraft of British design for the exclusive use of the American forces.

The Handley Page 0/400 had been selected as the main instrument of Trenchard's Independent Force operating against Germany from bases in Eastern

France and, in the absence of a satisfactory American heavy bomber, agreement was reached in December 1917 for provision of similar aircraft for a Night Bombardment Section to be set up as a component of the American Expeditionary Forces. These aircraft were to be powered by 350 hp Liberty 12-N engines and constructed initially in America, the components being shipped to the United Kingdom for assembly. The blockade of British ports by German submarines had drastically reduced the import of cotton, leaving both mills and operatives unemployed. Sir Herbert Dixon of the Cotton Control Board promised full support for the assembly of aircraft in disused Lancashire Mills and arrangements were made for the use of the Lilac Mill at Shaw and the Gorse Mill No 2 at Chadderton. Official records speak of the 'Oldham Aircraft Factory' and it is convenient to use the term but United States Army sources refer to the site as being in Failsworth and, in fact, parts of the site eventually developed fell also within the boundaries of Middleton and Manchester.

Although extensive work was done to alter the Lilac Mill for production, the war ended before this work was completed and no production was undertaken there. A large site, spreading from the Gorse Mill across the Rochdale Canal, was requisitioned to provide ground for an aerodrome and erecting sheds and London contractors, Trollope and Colls Ltd, were employed to build a large assembly shed close to the Mill. As a convenient site office the contractors used a large portion of the nearest hostelry, the 'Boat and Horses' and both the pub and the assembly shed remain to be seen today, the latter in the occupation of Her Majesty's Stationery Office.

American personnel were brought in to work on the aerodrome and erection sheds and a force of German prisoners of war set about the levelling of the field. Mill operatives were sent to Belle Vue to be trained in the rudiments of aircraft manufacture and the Alliance Manufacturing Company Ltd was appointed to run the factory.

Shipment of parts from America was erratic and, although 100 incomplete sets of components had arrived by the 25th September, 1918 and much had been done to prepare the aerodrome and its buildings, the Armistice on the 11th November brought all work to an immediate stop without a single Handley Page 0/400 having been delivered. The components which had arrived at the factory were repacked and returned to America and the local employees were discharged to find work elsewhere. The Oldham Aircraft Factory might well have produced significant numbers of effective aircraft if the war had continued and the adjacent aerodrome might have been put to use in post-war years. It is perhaps unfortunate that the project did not commence a little earlier.

There had been a certain amount of flying from the field adjacent to the works of A.V. Roe & Co Ltd and Trafford park had seen occasional use. Cringle Fields was to be used to some slight extent and plans envisaged extensive flying from the new airfield being constructed by the Americans but the Manchester area still had no aerodrome to compare with the larger fields in other parts of the country. In October 1917 the City Council were advised that the War Department proposed to build an Air Acceptance Park at Didsbury and the site chosen was to be beside the railway line between Chorlton and Heaton Mersey, stretching across what is now Hough End Playing Field to the vicinity of Old Hall Farm in Withington.

Hangars to take aircraft up to the size of the 0/400 were to be built and road works were required to improve the access to the area. The nearest railway station was Alexandra Park and it appears to have been from this that the aerodrome took its name as, although it was referred to as Didsbury in some documents, it was as Alexandra Park that it is generally recalled. The main military unit throughout the period when the aerodrome was under Service control was No 15 (Manchester) Air Acceptance Park and the wartime Commanding Officer was Major G. Williamson, MC, formerly of the Manchester Regiment. No 5 Works Training Squadron is also reported to have been attached.

Whilst the War Department supervised the erection of buildings, the laying of the aerodrome roads was undertaken by the City Engineer and to speed the work 50 Portuguese labourers were brought to Manchester. Not a great deal of information about the progress of construction has come to light but reports indicate that the ferrying of aircraft from the field was being undertaken by May 1918. A.V. Roe & Co Ltd took a hangar but Major Williamson is on record as stating that practically the whole of the output of No 15 A.A.P. would be drawn from No 2 National Aircraft Factory, approximately 300 aircraft per month being expected to arrive by road from Heaton Chapel. Certainly by the end of 1918 when all additional work on the aerodrome was cancelled by the Air Ministry, a number of large hangars were in use and the field was in a condition to receive the largest aircraft of the day. Both Service flying and manufacturers' test flying was taking place.

Whilst assembly work for the Americans at the Oldham Aircraft Factory ceased abruptly at the Armistice, the Avro works and the N.A.F. No 2 continued to produce aeroplanes. Manchester celebrated the end of the War and on the day after the Armistice, 12th November, 1918, the inaugural meeting of a new Manchester Branch of the Royal Aeronautical Society was held at the College of Technology and duly reported in the Avro house journal 'Joystick'. The Air Minister, Lord Weir, came to Heaton Chapel on the 20th December to receive a DH9 presented by the staff of the Aeronautical Inspection Department at the National Aircraft Factory and the aircraft was christened by the Lady Mayoress. Two days later it was announced that a number of large bombing aircraft from a camp in the south would fly to Alexandra Park and would give a display of coloured lights as well as providing a searchlight test. Regrettably, the weather was unfavourable and the flight did not take place.

When, four weary years before, Manchester's sons had marched away to war, very few had gone to join the Royal Flying Corps or Royal Naval Air Service but large numbers had transferred to those services or joined in succeeding years and they had been absorbed into the new Royal Air Force. When the time came for those who survived the War to return to their homes, many of them did so through No 1 Dispersal Unit which was to occupy a part of Heaton Park until May 1919 under the command of Lt Col E. Lascelles.

The end of the War saw Manchester men looking forward to the use of aircraft in the pursuits of peace and on the 13th December, 1918 the Manchester Guardian, by now convinced that the aeroplane had much to offer in both peace and war, advocated in its leader column that Manchester should become the main centre of flying outside London. On the following day the paper announced that a new company 'Manchester-London Air Service Ltd' had been formed. The prospectus stated that there would be a fleet of aircraft, each of 2,000 hp and carrying 20/40 passengers at over 100 mph giving a service at two-hourly intervals at a cost of less than sixpence per mile. Suitable aircraft and land for terminals were being sought.

The Company does not appear to have been successful but a new era was envisaged if not immediately to dawn.

The Struggling Twenties

The period from 1920 to 1930 might well be said to be divided in some aeronautical senses, at least as far as the Manchester area is concerned, into two quite separate parts related on the one hand to the running down of Service aviation after the war and the rise and fall of early attempts to provide for civil flying and, on the other hand, to the faltering steps taken in the second part of the decade towards the development of civil aviation and the efforts made to keep alive British military aviation in the face of apathy and impoverishment.

If the year 1919 does not fall into the category of 'the twenties' by any possible chronological calculation, it can equally not be construed as one of the years of the 1914-18 war and, as its activities in aviation fall closely into context with those of the immediately following years, it is convenient to consider it with this chapter.

The first months of 1919 saw some military aircraft being produced and Service flying continuing but the aircraft industry found that the large wartime orders had been very drastically reduced and hopes centred upon an early upsurge in the use of the aeroplane for civil purposes. Whilst some of the smaller aircraft constructing companies foundered, others sought new designs which might attract orders and held on for better days to come, in some cases turning over large parts of their factories to the building of buses and cars or the fabrication of any product for which a sale could be found. Large numbers of skilled workers found that their jobs evaporated overnight and design staffs were also reduced.

A.V. Roe and Co Ltd had the Miles Platting and Hamble Works with Newton Heath factory nearing the stage when it could be regarded as a major potential centre of production. Accommodation was held at Alexandra Park aerodrome alongside the Royal Air Force and the Company looked to the future by acquiring numbers of Avro 504s which it was felt would have a big part to play in educating the public to the adoption of the aeroplane as a familiar feature of everyday life. Several experiments were conducted to marry the proven airframe to both radial and in-line engines to take advantage of available stocks of surplus power plants and variants were produced with modifications to enable up to four passengers to be carried. The prototype Avro 536, K114, which flew in April 1919, was the first of these conversions to provide four seats behind the pilot and small-scale production was immediately undertaken. Another 504 variant was a version with a Sunbeam Dyak engine which was the progenitor of a number of similar aircraft converted in Australia, one of which was to become the first aircraft of the Queensland and Northern Territory Aerial Services Co Ltd, (QUANTAS). The Avro 548 Tourist had two passengers in tandem cockpits behind the pilot and the Avro 546 continued the 536 theme with 3 passengers being accommodated in an enclosed cabin. To produce a new aircraft utilising mainly 504 components, Roe and Chadwick designed the Type 547 triplane in 1919 and determined attempts to provide an Avro aircraft for the prestigious Schneider Trophy contest were frustrated when it was found that, instead of the intended 275 hp Rolls-Royce Falcon, the only available engine for the Type 539 seaplane was a lower powered Siddeley Puma and the Company's entry was relegated to reserve place with the British team.

Possibly the most interesting design evolved by Roy Chadwick in the months immediately after the War was the diminutive Avro Baby. The first aircraft powered by a Green engine of only 35 hp crashed on its first flight on the 30th April, 1919 but a second aircraft built from the wreckage flew within ten days and distinguished itself by winning a number of races before Chadwick, having recently been taught to fly, was seriously injured when he crashed it at Hamble. Rebuilt again, it was subsequently bought by Bert Hinkler, traded its British registration G-EAUG for Australian marks G-AUCQ and can now be seen in the Queensland Museum at Brisbane. Hinkler's Baby made epic flights from Croydon to Turin and later from Sydney to Bundaberg and other exploits of later machines of the same design included a flight to Moscow, service as a seal-spotter in Newfoundland and conversion to carry six narrow chord aerofoils in place of the upper wing for aerodynamic experiments.

However interesting these designs may have been, the number of aircraft produced was negligible when compared to the output of the war years and to keep the Company alive A.V. Roe and Co Ltd, in common with other large aircraft firms, diversified their output in an endeavour to retain at least some part of the skilled work force which had been built up. A small 12 hp car was designed and it was planned that production of this would occupy the whole of the Avro works in Manchester with the exception of some space for aircraft repair work, production of aircraft being undertaken solely at Hamble. A.V. Roe had built himself an enclosed motor cycle before the war to protect himself from the weather when travelling to his various flying grounds and he carried his inventive genius into the design of the Avro car, introducing a number of novel features including an opening roof and adjustable seats. Limited production was undertaken but this ceased in May 1920 when Crossley Motors Ltd acquired a controlling interest in A.V. Roe and Co Ltd and from then on the manufacture of Crossley car bodies joined the output of baby carriages and billiard tables in occupation of the buildings at Miles Platting and Newton Heath.

In the months which followed the end of the War, flying continued at Alexandra Park and as well as the products of the local factories other types of Royal Air Force aircraft came to the aerodrome. One of these, a Bristol F2b, crashed on the aerodrome on the 14th January killing a local pilot, Captain C.A. Brown, and a soldier from the Nell Lane Hospital whom he had allowed to join him for a test flight. Another aircraft crashed at Sparrowpit in Derbyshire a couple of days later and with the misfortunes of the DH10s also taking place at this time the area had more than its share of untoward incidents. One which proved expensive but which did not lead to any personal injuries occurred on the 18th January when the pilot of a DH9 flying from Berwick to Tadcaster became lost in foggy conditions and after passing over Manchester landed in a field at Mottram. Airmen from Alexandra Park were brought to guard the aircraft but shortly after midnight it was seen to be on fire. The aircraft was destroyed in circumstances which, according to the Ashton Reporter, 'were engaging the attention of the police' and a local farmer's son was subsequently committed to Chester Assizes charged with damaging the aircraft to the extent of £2,000 and stealing 30 gallons of petrol. However, it is fair to record that he was subsequently acquitted on both counts.

Manchester was given its first sight of one of the country's largest aircraft on the 24th January, 1919 when a four-engined Handley Page V1500 passed over the city on a flight from Belfast to Sheffield but civil flying was not permitted until the 1st May. Even before that date possible landing grounds were being sought. In

March the Altrincham Guardian reported that a site at Warrington had been selected for a 'flight-drome' for a proposed Liverpool-Manchester-Hull service and shortly afterwards the City Engineer was to include Blackley and Didsbury Golf Courses, Heaton Park and grazing land at Newton Heath in a list of possible landing grounds. In the next couple of years Fairfield Common, Buxton, Northenden, Macclesfield (Congleton Road), Knutsford, Bramhall, Wilmslow and Alderley Edge were all to be recorded as having actual or proposed landing grounds. Special permits were issued at Easter for civil flights in the close vicinity of certain aerodromes but no suitable ground was found in the Manchester area, although the possibility of flying from Belle Vue was considered and a member of the Jennison family was reported as stating that passenger flights would certainly be one of the attractions of the Gardens in future.

On the 1st May, Air Navigation Regulations, 1919, promulgated by the Air Ministry for control of civil flying, came into force and a number of inaugural flights were made. A few days before the Officer Commanding 15 A.A.P. at Alexandra Park had announced that accommodation had been set aside for civilian aircraft, although he had no instructions with regard to the supply of petrol or facilities for servicing.

The first civil aircraft to arrive at Alexandra Park was a Bristol F2b from Oxford piloted by a Captain West who brought a film 'The Warrior Strain' for the Market Street Picture House. On the following day West flew round the city dropping leaflets advertising the film. The weather on the 1st May was gloomy but cleared later in the day and in the evening the first transport flight from London to Manchester was completed when a converted Handley Page 0/400 arrived from

The Handley Page 0/400 which made the first transport flight from London to Manchester on 1st May 1919. *:British Airways*

Cricklewood piloted by Lt Col Sholto Douglas, MC — later to become Marshal of the Royal Air Force Lord Douglas of Kirtleside and to include amongst his appointments the leadership of Fighter Command of the Royal Air Force and the Chairmanship of British European Airways.

Douglas had become Chief Pilot of Handley Page Ltd after his demobilisation from the RAF and received commercial licence No 4 on the 1st May. After collecting his licence he went to Cricklewood and took off with 11 passengers on the first leg of a tour arranged for the Daily Mail. He landed at Alexandra Park after a flight of 3 hrs 40 minutes. It was planned to fly on early next morning but after a short flight Douglas found that the weather was unfit and landed again. During the day he took the Deputy Mayor and newspaper reporters for local flights and made preparations to leave for Aberdeen on the following day.

On the 3rd May the 0/400 was wheeled out and narrowly escaped disaster when a DH9A piloted by F/Sgt. R.T. Parry, AFM stalled and crashed close by. Major Orde-Lees, a parachute expert who had accompanied Douglas was burned on attempting to rescue the pilot who died later in hospital. Orde-Lees had been brought on the flight to assist in the delivery of bundles of copies of the Daily Mail which were to be dropped on golf courses along the route and he appears to have been fit enough to join the aircraft when it left later in the day and reached the vicinity of Burnley before being forced back by the weather. Douglas eventually made the flight to Scotland on the 5th, dropped Orde-Lees by parachute at Aberdeen and landed at Turnhouse, Edinburgh. Another 0/400 piloted by Captain W.H.N. Shakespeare, MC, DFC made a further flight to Scotland on the 8th, both aircraft throwing out copies of the Daily Mail as they went.

Douglas was back at Alexandra Park for a display on the 10th May when pleasure flights were available at £2 each and Orde-Lees' assistant, Miss Sylvia Boydell, parachuted from the aircraft to land in a ploughed field at the junction of Wilbraham Road and Princess Road South. Two days later she was injured along with Orde-Lees and Shakespeare when the aircraft in which they were making another newpaper flight from Manchester to Scotland landed at Harker near Carlisle with engine trouble and crashed on an attempted take-off.

As well as diversifying into the production of non-aeronautical items, A.V. Roe and Co Ltd sought other ways of attracting income to tide the Company over the difficult period after the War and it was realised that not only were large numbers of Avro aircraft surplus to Service requirements but, with their aeroplanes, many competent pilots were no longer required by the Royal Air Force. The stories of their wartime achievements had given rise to a new interest in flying amongst the general public and there appeared to be a ready market for pleasure flights.

It was not possible to find an available flying ground near Manchester at Easter but three Avro 504Ks were used to provide joy rides at Hamble and the public eagerly seized the opportunity of taking to the air. The Avro Transport Company was formed and joy-riding centres were set up during the summer of 1919 at Alexandra Park which was the home base of the large fleet and also at over a dozen resorts in England and Wales. John Lord attended a meeting of the Blackpool Council on the 6th May and obtained authority to use a site opposite the Star Inn as the terminus of a Manchester-Southport-Blackpool service and a week later a party of officials from Blackpool was shown round Alexandra Park and then flown home. On the 26th May the service was inaugurated when two aircraft left Alexandra Park and the first scheduled domestic air service in Britain had commenced. It was to run for 18½ weeks without suffering a single forced landing, although occasional flights were cancelled due to weather conditions. The service was run without government subsidy and the aircraft left Blackpool at 12.00 noon and Southport at 12.15, arriving at Alexandra Park at 12.45. The return journey commenced at 2.00 pm and fares were 5 guineas single and 9 guineas return.

Throughout the summer the aircraft of the Avro Transport Company flew from beaches and fields throughout the country. As well as the aircraft used at the major resorts, other machines were taken to fields near to many towns to stay for a few days giving flights and floatplane versions of the 504 were used to fly passengers from Windermere. At Blackpool alone over 10,000 people were flown. When the season was over and the Manchester-Blackpool service closed down, the aircraft were brought back to Alexandra Park where the Chief Engineer, J.C.C. Taylor, who had spent the summer trouble-shooting round the various sites in the sole Avro 538, K132, set about overhauling them for the following year.

The advent of civil flying brought many people to see the aircraft but not all the visitors to the aerodrome were welcome. Trouble was experienced with people trespassing both by day and apparently also by night. Several appeared in the City Magistrates' Court and on the 16th May the Evening Chronicle reported that the airmen had had great fun during the previous night by using Verey pistols to illuminate and apprehend courting couples on the aerodrome.

Whilst the civilians and the airmen pursued their respective sports at the aerodrome, numbers of people in the centre of the city visited an exhibition at Finnigans' store on Deansgate where a Norman Thompson flying boat had been placed on display.

Aircraft continued to expand their journeys about Britain but on the 15th June, 1919 two Manchester men made a flight which presaged the extension of commercial flying from internal to not only international but inter-continental routes. One of them, John Alcock, had already established himself as a pilot of some repute when war broke out in 1914 and he spent the first year of the war testing aircraft and instructing the growing number of pilots required for operational flying. His constant pressure for more active participation in the war eventually led to his posting to No 2 Wing of the Royal Naval Air Service where he put his experience to good purpose both in bombing raids in a Handley Page 0/100 and also in the contrasting role of fighter pilot being credited with at least two enemy aircraft whilst flying a Sopwith Camel. Alcock's operational career ended abruptly when he was forced down whilst flying the 0/100 and he became a prisoner of war.

Arthur Whitten Brown, although born in Glasgow, had moved to Chorlton at an early age with his parents and trained as an engineer. He was commissioned in the Manchester Regiment and fought at Ypres before being transferred to the Royal Flying Corps as an observer. Shot down over enemy-held territory, he was injured in the left leg and taken prisoner. For almost two years Brown languished in prison camps and he spent much of his time studying navigation. This was to stand him in good stead when, soon after his release, he visited the Vickers works at Weybridge and was introduced to John Alcock who had undertaken to pilot the Vickers Vimy which had been nominated as a contestant for the £10,000 prize offered by the Daily Mail in 1913 for the first non-stop Transatlantic Flight.

On the morning of the 14th June, 1919 the Vimy which had crossed the Atlantic in crates for assembly in Newfoundland took off to cross almost 2,000 miles of ocean before landing in an Irish bog, having made the first-ever direct

Transatlantic Flight. Alcock and Brown were taken to London and were knighted by King George V. A month later they flew to Alexandra Park and, after driving in an open carriage past cheering crowds to the Town Hall, each was presented with an illuminated address by the Lord Mayor. Medals were struck for presentation to them but tragically Alcock was not to survive to receive his. When the furore of the Transatlantic flight had died down he returned to test and delivery flying with Vickers and on the 18th December he was killed when a Vickers Viking amphibian crashed near Rouen in France. After being brought to Manchester for a service at the Cathedral, he was buried in Southern Cemetery as Avro aircraft from the nearby aerodrome dipped in salute.

Whitten Brown returned to engineering and died in Swansea in 1948, his memory with that of Alcock perpetuated by the citizens of Manchester in memorials in the Town Hall and at the present airport where a collection of relics of the historic flight is displayed to remind the thousands of passengers who cross the Atlantic of the two Manchester men who flew from America in the open cockpit of the Vimy.

Whilst civil flying attracted many of the pilots who had survived the war and others returned to their jobs in Manchester and its surrounding towns, the city and the nation were reminded that the cost of winning had not been in human sacrifice alone. The public were exhorted to contribute to a great Victory Loan and in July 1919 the campaign was furthered in Manchester by the exhibition of an Avro 504 in Albert Square whilst similar aircraft flew over the district dropping leaflets. Investors in Albert Square were rewarded with small fragments of shot down enemy aircraft and on the 9th July the airship R33 flew over the city in support of the Loan.

The first days of October 1919 found Britain in the grip of a rail strike and the Government organised an air mail service between London and the major cities. A number of demobilised pilots were recalled to the Royal Air Force and DH9A aircraft were used, mail being flown between London and Manchester via Castle Bromwich and Shotwick and distributed throughout the North of England by aircraft of the 'Manchester Circuit' under the command of Captain G.J. Monson-Fitzjohn who was described as the O.C. Manchester Storage Park, RAF. One of the aircraft landed in a field beside Parsonage Road, Heaton Moor on the 6th October and the pilot, Lt Coleman, is reported to have delighted a large crowd by performing a loop when he took off the following day.

If 1919 had been a year of hope in aviation, the immediately succeeding years brought little to sustain that hope. The existence of large numbers of wartime aircraft and the fervent desire of both the Government and the population to believe that they had really seen a war to end wars minimised orders for new military aircraft and with the prevailing financial stringency the presence of the same surplus aircraft gave little encouragement to the production of machines for a commercial world unable or unready to support civil flying. The DH9s and DH10s had ceased to flow from the factory at Heaton Chapel and Cringle Fields reverted to the Parks Department. The training establishment at Belle Vue became a great indoor stadium and the Avro factories found little work.

The fleet of aircraft which had carried so many passengers in 1919 passed the winter at Alexandra Park but it became apparent that the people who had flocked to fly as an expression of post-war release were finding that work and funds were now harder to come by and in April 1920 the decision was taken to cancel joyriding plans and move the aircraft to Newton Heath which became the 'Head Office of A.V. Roe & Co Ltd in August 1920. Several aircraft were to be

The first production Avro 549 Aldershot exhibited at Alexandra Park during the return leg of the King's Cup Air Race 9th September 1922. *:E.A. Harlin*

acquired by individual pilots and small concerns to carry on financially precarious barnstorming tours. Some comfort was obtained from an order to recondition Avro 504s for the Royal Air Force and limited numbers of similar aircraft were delivered for both civil and military flying in overseas countries as the fame of the 504 as a training aircraft spread world wide. Over 600 aircraft to the design were built in Russia and many other present day air forces were also founded on either surplus RAF machines or new-built 504s.

In September Roy Chadwick commenced the design of a massive single engined bomber the Avro Aldershot, with a 68 ft span and power derived from a 650 hp Rolls Royce Condor engine. Prototypes were built at Hamble and one was later flown by Bert Hinkler with a 1,000 hp Napier Cub but it was for a version with the Condor that a production order for 15 aircraft was eventually received, the aircraft reaching No 99 Squadron at Bircham Newton in 1924. The Aldershot marked the first step in a gradual change from wooden to metal construction in the Company and for the first time fitters were used in the building of fuselage structures.

There was little flying at Alexandra Park and by June 1920 the aerodrome was deleted from the list of Service airfields and recorded as a licensed civil aerodrome without facilities. Hangars were used as billets for Territorial Army soldiers and by November pilots were warned that grazing rights had been let and that they might expect to find cattle and sheep on the aerodrome. In the following month work commenced on the extension of Princess Road to the south curtailing the area available for flying, although adequate space remained. Occasional flights continued and in June 1921 pictures of the Derby were flown to Manchester for Northern newspapers. Daily Mail pictures of a royal visit to Belfast arrived in the same month in the DH4A, G-EAMV of Captain F.L. Barnard and a DH9 brought Alan Cobham who had earlier visited the area giving pleasure flights at Bolton and Warrington. An R.T. station was maintained at the aerodrome using the call sign GEM until November 1921.

Whilst these were dark days for aviation, there were still some who looked ahead to better days and in April 1922 the Manchester Aviation Co Ltd which had been formed a few months earlier applied to the Air Ministry for the use of Alexandra Park for services to Dublin and to various points in the United Kingdom, announcing that if this permission was not forthcoming services would be flown from an aerodrome at Hazel Grove. The Company organised a display on the 29th April at which the Chief Constable, Sir Robert Peacock, made the first flight in a DH6 and the public clamoured for flights in the three aircraft which were available. The proposed services do not appear to have commenced and, although the Company acquired a number of Avro 504s which were used extensively for joy-riding and also a DH9C,G-EBDG, it finally went into liquidation in 1924, the enforced sale of four aircraft and a number of spare engines fetching only £600.

The Corporation was informed in June 1922 that the Didsbury aerodrome had been declared surplus to RAF requirements and Major General Sir Sefton Brancker, Director of Civil Aviation, discussed with members of the City Council the possible purchase of the aerodrome. The Lord Mayor declared himself in favour and he was supported by the Manchester Guardian in its leader of the 27th September. Negotiations continued into 1923 and it seems likely that Manchester would have had a municipal aerodrome at this time if the owner of the land, Lord Egerton of Tatton, had not refused to sell the area to the City for flying purposes when the Government's lease expired.

The presentation of a cup by King George V to encourage sporting flying attracted 21 starters for the first King's Cup air race on the 8th September, 1922, the course for the first day being Croydon-Birmingham-Newcastle-Glasgow with stages via Manchester and Bristol back to Croydon on the following day. The race was well supported by Manchester and the City Council provided a prize of 50 guineas for the fastest handicap time from Croydon to Alexandra Park, with the Manchester Guardian and Mr William Kayly giving prizes for the fastest and second fastest aircraft on the leg from Glasgow. John Lord of A.V. Roe and Co Ltd was in charge of arrangements at the aerodrome and RAF aircraft from Shotwick participated in a flying display. Thirteen aircraft left Glasgow and Fred Raynham in his bright yellow Martinsyde F6, G-EBDK and Alan Cobham in the DH9B, G-EAAC arrived together followed very shortly afterwards by Barnard, ultimate winner of the race in the blue and silver DH4A, G-EAMU *City of York*. Two aircraft, the Armstrong Whitworth Siskin, G-EBEU and the Avro Baby, G-EAUM were unable to continue from Manchester but the remaining 11 aircraft reached Croydon safely, Raynham and Cobham taking second and third places behind Barnard.

On 1st October, 1922 the Government announced its subsidy scheme for routes to be operated by British airlines and £55,000 was awarded to Daimler Hire Ltd for a Daimler Airways service from Manchester via London to Amsterdam with a connection to Berlin. The aircraft to be used were de Havilland DH34s powered with a single 450 hp Napier Lion engine and painted overall in bright red. Nine passengers were carried in an enclosed cabin behind the open cockpit and the fare from Manchester to London was £2. 3s. 0d. — just 1½d. more than the cost of a first-class rail ticket!

Crowds gathered at Alexandra Park on the 23rd October to see Captain G.R. Hinchliffe take off in G-EBBS to inaugurate the service, meeting the first aircraft from Amsterdam to Manchester at Croydon. Scheduled time for the flight to Amsterdam was four hours, lunch being taken at Croydon and a call being made at Rotterdam on the way. By the 12th April, 1923 the Manchester Guardian was able to announce that arrangements had been made for through bookings from Manchester to Moscow to be made. On the 30th April the first aircraft to make a through journey from Manchester to Berlin left Alexandra Park and reached the German capital before sunset. Proposals for a service to Zurich were shelved in the face of limited support and a suggestion that the London route should be routed via Norwich does not appear to have found favour.

Only two incidents marred the Daimler service. On the 20th August one of the aircraft was forced to land in a field at Wigginton near Tamworth but took off safely next day and on the 14th September the aircraft which had flown the inaugural flights to both London and Berlin, G-EBBS, crashed near Ivinghoe Beacon on the north edge of the Chiltern Hills whilst on the evening flight from London, killing the three passengers and both pilots, G.E. Pratt and L.G. Robinson. For a time the service was suspended but further regular flights were made in 1924 before the Daimler fleet was taken over by Imperial Airways Ltd who did not include Manchester on their schedules.

Experimental flights were made in 1923 to establish an air mail service between Plymouth, Birmingham, Manchester and Belfast, the Mayor of Plymouth arriving with one flight in September and advocating that the service would have the great virtue of enabling Devonshire cream to be airfreighted to Manchester along with His Majesty's mails! Although the de Havilland Company and their pilots, Alan Cobham and Hubert Broad, made strenuous efforts to find support for the service, it did not progress beyond the experimental stage.

The King's Cup Race for 1923 again included Manchester as a control point and on the 14th July the Royal Air Force once more provided support for a display to coincide with the arrival of the racing aircraft. By now the partially completed Princess Road formed an embankment as it rose to pass over the railway line beside the aerodrome and many spectators preferred to view the activities from there rather than pay to enter the field. Nine aircraft reached Alexandra Park from Glasgow, the Manchester Guardian prize for the fastest coverage of that leg being won by L.L. Carter in a Gloster Grebe, G-EBHA and there was a sensation when the previous year's winner, Captain F.L. Barnard, was disqualified for failing to cross the designated line. Captain H.C. Biard, winner of the Schneider Trophy in 1922, passed through in the Supermarine Sea Eagle, G-EBFK, although out of the race. The eventual winner was Captain F.T. Courtney in the Siskin, G-EBEU with which he had been compelled to drop out of the race at Manchester the previous year.

In 1924 Manchester was not included in the King's Cup course. The regular service to London ceased and as the time for the expiry of the Air Ministry's lease of the aerodrome approached, A.V. Roe and Co Ltd sought land for a new airfield for the testing and delivery of their aircraft. At the end of August all flying ceased, on the 17th September there was an auction of remaining effects and the 30th November was set as the final deadline for the removal of all hangars and buildings. By the 18th November the Manchester

Guardian reported that three of the hangars were being put up on the land at New Hall Farm which was to become the new Avro aerodrome at Woodford but in the event only one was erected.

When A.V. Roe and Co Ltd were forced to seek — quite literally — fresh fields and pastures new with the closure of the Alexandra Park aerodrome they were not alone. For some time the aerodrome had been the home of a small glider owned by the Lancashire Aero Club which had been stemmed from the efforts of a man who was to leave his considerable mark on a number of aspects of Manchester's civil aviation history. John F. Leeming was born in Chorlton and commenced to take an interest in aviation in the pre First World War days of his youth in Southport where he had followed the construction of models in various configurations with the construction in 1910 of a 20 ft span monoplane and three other gliders. None of these appears to have been strikingly successful but in 1922 Leeming still cherished an ambition to create an aircraft and, spurred on by reports of the feats of German glider pilots, he set about designing a glider which would not only provide a means of flying but which could be dismantled and stored in a small garage. Aware of the large quantities of surplus material held by some of the aircraft manufacturers, John Leeming, who had by then come to live in Hale, contacted the nearest factory to his home and was received by John Lord who was sufficiently impressed by his enthusiasm to give him the run of the Avro scrap-heap. At Newton Heath Leeming was introduced to Clement Wood and Tom Prince and they quickly joined in the project, being followed shortly afterwards by a fitter, George Bayliss and a young businessman, Mark Lacayo. The glider was designated the LPW, taking the initials of the three founders of the small group and construction progressed — more slowly than some of the enthusiastic young men would have wished — under the meticulous supervision of Tom Prince who combined his authority as an Avro foreman with a strict regard for standards of accuracy and care in his spare time work. He was joined in this by a colleague from Newton Heath, Arthur Ainsworth, who many years later was to supervise the rebuilding of one of the last Avro 504s which is still flying with the Shuttleworth Collection. The glider was moved from the garage of Leeming's home into the greenhouse and it was in the greenhouse that the Lancashire Aero Club was formed with ten enthusiastic members. It was not the first club to bear the name as the Blackpool flying meetings of 1909 and 1910 had been organised by a Lancashire Aero Club which did not survive the war. In 1919 efforts had been made to form a Lancashire Flying Club at Southport with help from pilots of the Avro Transport Company but little seems to have come of it and an attempt to give that title to the new club was firmly rejected by Tom Prince who pointed out that there were already many flying clubs — nearly all associated with pigeon racing!

The glider was moved to Alexandra Park and made its first flight on the 24th May, 1924, being towed into the air at the end of a long rope attached to a car. The Lancashire Aero Club progressed apace, Lord Leverhulme accepting the Presidency, and applied for affiliation to the Royal Aero Club but the closure of Alexandra Park provided a set-back and negotiations took place with a view to obtaining a new flying ground at West Timperley. However, A.V. Roe and Co Ltd once again came to the club's aid and permitted the use of their new aerodrome at Woodford where the LPW, fitted later with a Douglas engine, made continuing efforts to take to the air. Never conspicuously successful either as a glider or with its diminutive engine, it was nonetheless the vehicle around which the first light aircraft club in the country was formed and the forerunner of a long

John Leeming taking off in the LPW glider at Alexandra Park. :*B.A.G. Meads*

line of aircraft owned by the Club to the present day.

When in late 1924 the Air Ministry decided to subsidise the formation of a number of clubs, the Lancashire Aero Club was already in being and on the 18th November Alderman Sir William Kay presided at a meeting at the Midland Hotel where Leeming was able to announce the donation of an Avro 504 to the Club. Under the Government's scheme two de Havilland Moths were acquired, G-EBLR and G-EBLV, the latter of which is still flying more than fifty years after its delivery to Woodford, and to celebrate the delivery of the second aircraft on the 29th August, 1925, a pageant was organised at the aerodrome. Vast crowds were being attracted to the annual displays at Hendon but the general public in other parts of the country had few opportunities of seeing aircraft at close quarters. The Lancashire Aero Club was the first of its kind to perceive that many people would pay to see flying and some of them might well be impelled to seek flying instruction and membership of the Club. A month later a more ambitious event saw 25,000 people at Woodford for a display which included Bristol Fighters and Sopwith Snipes of the Royal Air Force as well as numerous performances by well-known pilots and a race which attracted 18 entries. Turning points were at the Goyt Mill, Marple and the Knutsford Water Tower and the winner was F/0 H.R.D. Waghorn, later to represent Great Britain in the Schneider Trophy races. By the beginning of 1926 the Club supplemented its two voluntary instructors, J.C. Cantrill and J.J.(Joe) Scholes, by engaging its first full-time instructor, Neville Stack, and obtained two new aircraft, the Moth, G-EBMQ presented by Sir Charles Wakefield, newly appointed Club President, and the Avro Gosport, G-EBNF, presented by A.V. Roe and Co. Ltd.

The mid-twenties saw some effort by Imperial Airways to establish passenger flying but this activity did not impinge on Manchester, nor did the area see much of the work which Trenchard was doing to re-organise the Royal Air Force. Apart from the gradual build-up of the Lancashire Aero Club and the struggles of A.V. Roe and Co Ltd to keep the aircraft industry alive in the North West, there was little to be seen of aviation in one of the country's greatest industrial regions. From time to time an unfortunate landing hit the headlines and on the 11th May, 1925 the pilot of a Bristol Fighter from Sealand became lost over Rochdale and attempted to land at Greave where he struck a tree and was extracted from the remains of his aircraft by players of Rochdale Football Club.

At Newton Heath, acceptance of the Avro Bison following trials at Martlesham Heath in 1922 led to the ultimate production of 41 of these large Fleet Spotter biplanes for use on *HMS Furious* and *Eagle* as well as at various shore stations. In addition, Roy Chadwick designed a twin engined aircraft of 96 ft span powered by 2 x 650 hp Rolls-Royce Condor engines. This aircraft, the Type 557 Ava, carried a crew of five but did not progress beyond the construction of two prototypes. Aldershot parts were used in the construction of the Avro Andover which was a Condor-engined transport/air ambulance type of which four aircraft serialled J7261 - J7264 were ordered for the RAF in January 1924. Attempts were made to win contracts with the prototypes of the Avenger, a Lion-engined single seat fighter, and the deck-landing Buffalo with a similar engine but neither progressed beyond the prototype stage. The Lynx-engined Avocet floatplane did not achieve production orders, although one Avocet was used as a

The second of the Lancashire Aero Club's original DH Moths at Woodford. The people are believed to be: L-R Sir Charles Wakefield, John Leeming, Sir Samuel Hoare, unknown, unknown, Alan Cobham. *:B.A.G. Meads*

practice aircraft for the 1929 Schneider Trophy team.

In 1926 the Air Ministry began to take an interest in the possible advantages of rotary winged flight and ordered a series of Avro produced autogyros to the designs of the Spaniard Juan de la Cierva.

Although every attempt was made to obtain Service orders, it was for the civil field that an Avro aircraft was next to be produced in numbers. Completed for the 1926 light aeroplane trials, the Avian prototype G-EBOV won several races before it was bought by Bert Hinkler and used for a 1200 miles non-stop flight to Riga in Latvia. With an 85 hp Cirrus II engine replacing its original Genet and a new undercarriage and wings, the aircraft was flown home to Australia by its owner and is now preserved in his honour in the Queensland Museum at Brisbane where Hinkler's Avro Baby is also kept. Production of the Avian commenced in 1927 and large numbers contested the lightplane market with the de Havilland Moth. Several were used for record-breaking long distance flights all over the world and the number of different engines fitted reached double figures. A few survive to the present day including G-EBZM owned by the Aeroplane Collection Ltd which is displayed at the Torbay Aircraft Museum and G-ABLK which, after crashing in the Sahara Desert, lay there from 1929 until being retrieved for rebuilding in 1976.

In 1928 A.V. Roe and Co obtained a licence to build Fokker F.VII three-engined transports and fourteen aircraft were produced of which five were sold to the first Australian National Airlines Ltd. Seven scaled down variants were also produced and a larger design, the Avro 642, with a similar wing and either two or four engines was also built.

Gradually the paucity of orders which had followed the First World War was receding and flying was beginning to attract the interest of the masses as more aircraft were seen about. In the later years of the nineteen twenties Manchester people read more of aviation in their papers and a gradually growing proportion of what they read related to their own area. Those two great protagonists of civil aviation Sir Sefton Brancker and Sir Alan Cobham were strong advocates of both the development of club flying and the opening of new aerodromes for the use of civil air transport and their views were echoed locally by John Leeming. Within a week of landing dramatically on the Thames at Westminster after a round trip to Australia in a DH50 for which he was knighted by King George V, Cobham flew to Manchester, landing at Woodford. He was flown from there to a hastily prepared landing ground at Old Hall Farm, Withington, during the afternoon of the 7th October, 1926 in an aircraft of the Lancashire Aero Club piloted by Leeming with the two Moths flown by Cantrill and Scholes formating as escorts. The City was celebrating the success of its Civic Week and Cobham was enthusiastically received by the Lord Mayor, Alderman Miles Mitchell, and a large crowd whom he addressed from the Town Hall steps. Here — and later at the University where he was made an Honorary Life Member of the Union — he advocated the setting up of more aerodromes for civil aviation, a call which was to find growing support in Manchester in the coming years.

Cobham's overseas flights had done much to set the public and the business world thinking and talking about flying and, as Director of Civil Aviation, Sir Sefton Brancker, sought every opportunity of furthering this process. In conversation during one of Leeming's visits to London he casually dropped the suggestion that excellent publicity would accrue from the successful landing and subsequent take-off of a light aircraft from the top of one of Britain's better-known mountains. Leeming at once took up the idea and immediately saw the possibility of linking publicity for Manchester aviation to publicity for a Manchester aircraft. After some misgivings on the part of the manufacturers, it was agreed that a lightened Avro Gosport could be made available and that the chosen aircraft would be fitted with the new experimental 90 hp Avro Alpha radial engine. First plans were for a landing at the summit of Snowdon but it was decided that conditions there were such that even the sturdy Gosport was unlikely to survive the attempt and Helvellyn was chosen. The Works Manager, Roy Dobson, set about the preparation of the aircraft, registered G-EBPH, and it was agreed that Bert Hinkler would accompany Leeming in another aircraft. After the failure of a preliminary attempt due to adverse weather conditions, the landing was finally achieved on the 22nd December, 1926 and, in the event, Hinkler flew with Leeming in the Gosport as the aircraft which he had intended to use was unserviceable. To Leeming's surprise, he found on the summit a Professor of Greek who readily agreed to sign a scribbled certificate to provide evidence of the landing. A successful take-off was made over Striding Edge.

Shortly before Leeming and Hinkler made their visit to Helvellyn, Captain T.N. Stack, the Lancashire Aero Club's paid instructor and B.M.T.S. Leete, a member of the Club who had seen service in the Royal Air Force, set off for an adventurous flight which took them in two DH Moths from Stag Lane via Paris, Malta, Benghazi and Baghdad to Karachi. Although the two pilots described the venture as a holiday jaunt, their 5,500 mile flight was regarded as sufficiently meritorious to justify the award of the Air Force Cross and a number of Indian flying clubs were set up as a direct result of the publicity given to their arrival.

In 1927 Woodford saw growing numbers of Avro Avians leave the assembly

The Avro 585 Gosport returning from Helvelyn on 22nd December 1926.
:Manchester Guardian

sheds for delivery and during the year three notable flights were made from the aerodrome by a 22 year old Irishwoman, Mrs Sophie Elliott-Lynn, who later became Lady Heath. On the 19th July she took off alone in the early morning and succeeded in landing at no less than 79 places in a tour of aerodromes and flying fields which commenced at a field near Altrincham and ended in the late evening at Newcastle-on-Tyne. This flight was made in the Lancashire Aero Club's Avian, G-EBQL and was followed on the 25th July by another flight in which the aviatrix succeeded in visiting four countries in the British Isles in a single day — no mean feat having regard to the 82 mph cruising speed of the Avian. Mrs Elliott-Lynn made a number of flights in Europe during the summer but returned to Woodford on the 8th October when she succeeded in reaching a height of 19,200 feet in the Alpha engined Avian II, G-EBSD with a passenger.

Flying was rapidly gaining in popularity as the clubs grew in strength and, with the growing numbers of private owners, eagerly acquired the Avians, Moths and other light aircraft which were being produced for the increasing market. Inevitably, not all of the stories which the newspapers carried had the happy endings of the successive record-breaking flights which were reported. The area around Manchester had its share of mishaps, although most were of only minor significance. An RAF aircraft from Sealand made a forced landing in a field off Boat Lane, Diggle on the 28th March, 1927 and turned over, leaving the pilot, F/O Hare, hanging upside down until he was released by local inhabitants. Rather more fortunate were the occupants of an Avro 536 of Berkshire Aviation Tours Ltd who were thrown from the aircraft but unhurt when, during a display at a field beside Chester Road, Altrinchan on the 25th May, 1928, the pilot failed to see a line of telephone wires and was brought to a rapid and unexpected halt. The Avro Gosport, G-EBNF was written off with fatal consequences when it crashed on No 7 green of Bramhall Park Golf Course on the 21st October, 1928, the subsequent Air Ministry enquiry deciding that the passenger had unwittingly interfered with the controls whilst the aircraft was in a spin.

The Lancashire Aero Club had its share of relatively minor mishaps but the Moths and Avians were usually repaired and in the air again within a short time, thanks to the skill and perseverance of Walter Bartram and the other ground engineers at the Club.

Interest in flying had developed as Britain drew away from the years of war and the Lancashire Aero Club gave opportunities for keen enthusiasts to learn the art. Pleasure flights could be had by daring members of the public with some of the barnstorming pilots who still toured the country or set up at the North West holiday resorts, and the order books of the aircraft manufacturers were beginning to take on more healthy appearance, but for the businessman or traveller, the Manchester area — in common with most of the country — offered little in the way of aerial transport. Imperial Airways and the Royal Air Force pioneered the establishment of international air routes and as early as March 1920 Croydon had been established as the London Customs Airport.

It is possible that Manchester might have been even earlier in the field of municipal airport operation if Lord Egerton of Tatton who owned the site of the Alexandra Park aerodrome had been prepared to lease or sell the land but he was adamant that it should not again be used for flying purposes and, although the area passed to the City, it was as Hough End Playing Fields that it became known, only the Headquarters of 184 Squadron of the Air Training Corps today providing a tenuous link with the Royal Air Force's occupation. As early as August 1926 John Leeming had written to the Manchester Guardian advocating the establishment of an aerodrome in the area and Brooklands had been suggested as a site along with a small parish named Ringway to the south of Wythenshawe. Cobham's visit kept alive an interest amongst the more far-sighted members of the City Council and the Chamber of Commerce and, in 1928, following Leeming's publication and distribution of a booklet on 'Manchester and Aviation,' an Aerodrome Sub Committee was formed. Manchester was not alone in the area in considering the possibility of setting aside land for an aerodrome and in early 1928 both Bolton and Salford were pressed to have their own fields but by June the City Surveyor called upon Leeming to report on possible sites for a City aerodrome. John Lord wrote indicating that A.V. Roe & Co Ltd might well consider closing Woodford and moving to a suitable municipal aerodrome and in October the Sub Committee inspected a number of proposed sites. On the 17th October, 1928 the Manchester Evening News came out with the headline 'Ideal Site Selected at Barton' and announced that Manchester's new aerodrome would be within six miles of the heart of the city and even nearer to the great industrial estate at Trafford Park.

Having decided to become the first city in Great Britain to have a civic aerodrome, the Council set about preparing the site at Chat Moss but it soon became apparent that, although later with a relevant decision, Hull was likely to have a suitable ground in operation at Hedon. Little prestige ever accrues to the runner-up in a race and to ensure Manchester's precedence in the field, a temporary site was found at Wythenshawe, where the Rackhouse estate now stands. Hedges were removed and a barn became the terminal building and hangar.

To be successful, an aerodrome, whatever the form of its management, must have aircraft and operations and Leeming's concern that Manchester should not be left with a field in only limited use led him to consider the establishment of a company to provide air taxi and other services. In late 1928 he resigned his Chairmanship of the Lancashire Aero Club and joined with F.J.V. Holmes of Berkshire Aviation Tours Ltd in the formation of Northern Air Transport Ltd and Northern Air Lines (Manchester) Ltd. A DH9 was purchased and a Moth and two Avians were added to the large fleet of Avro 504s brought into the concern by Holmes. Captain A.N. Kingwill, a greatly experienced veteran of the Royal Flying Corps, was engaged as Chief Pilot and he was to prove a great asset in the difficult years of the enterprise's existence. Forty-five years after making the first landing at a municipal airport in Britain in the Moth, G-EBZU, he was to renew his acquaintance with Manchester aviation at a ceremony to mark the opening of extensions to the Terminal at the present Manchester Airport.

The initial landings at Wythenshawe were made on the 2nd April, 1929 and Kingwill was followed by Miss Winifred Brown in her Avian, G-EBVZ and J.C. Cantrill, Peter Eckersley (later to become Captain of the Lancashire County Cricket Club) and Alan Goodfellow in similar aircraft of the Lancashire Aero Club. On the 29th April The Lord Mayor (Alderman George Westcott) flew to London in the DH50, G-EBQI *The Lancastrian* of Northern Air Lines to collect from the Air Minister Aerodrome Licence No 1414 dated 19th April and valid for an initial period of six months from the 22nd April. The DH50 went unserviceable at Croydon and the Lord Mayor returned to Wythenshawe in the DH9C, G-EBIG to brandish an entirely irrelevant piece of paper to the waiting crowd, the licence having been inadvertently left in London!

Wythenshawe did not remain open for a long period but a number of interesting visitors used the field as well as the resident aircraft of Northern Air

Captain A.N. Kingwill (pilot) being greeted by Councillor Davy in front of the DH Moth in which he made the first landing at Wythenshawe 2nd April 1929. John Leeming stands on Capt. Kingwill's left. :Manchester Guardian

Lines and the frequent training flights of the Lancashire Aero Club. The first foreign-built aircraft was the Fokker Universal, G-EBUT, the first Service aircraft a Gipsy Moth from Sealand and the largest visitor the Ford Trimotor, NC8412 which called on a demonstration flight. It is interesting to observe that at least two of the visitors can still be seen, the more famous of them being the Gipsy Moth,G-AAAH *Jason* which is in the National Aeronautical Collection, honouring its later owner Amy Johnson. At present under restoration is the Avian, G-ACGT which was initially registered EI-AAB and the second civil aircraft registered in Eire.

Although John Lord had shown a willingness to discuss a possible transfer of A.V. Roe & Co's test flying from Woodford to a new Manchester aerodrome, nothing came of the suggestion and Lord himself was to leave the Company shortly afterwards. When Alliott Verdon-Roe sold out his interest in the growing organisation which had taken his name, John Lord left with him and together they took over the firm of S.E. Saunders Ltd and formed Saunders-Roe Ltd. From 1928 A.V. Roe & Co Ltd came under the control of Sir W.G. Armstrong Whitworth Aircraft Ltd and the end of the decade saw the production facilities well occupied with 504Ns for the RAF and foreign Services as well as an encouraging flow of Avians for the civil market.

The Awakening Thirties

The years following the end of the First World War had seen aviation in Great Britain make a number of faltering steps in the face of a continuing lack of major government support and owing much to a few men of vision in both civil and military fields. Some steady progress had been made and as the manufacturing concerns produced successive improvements in airliner and warplane design they were encouraged by the spread of interest in lightplane flying amongst private owners, club flyers and the many members of the public who flocked to see aeroplanes when displays enabled them to do so. As the nineteen twenties passed into history, there was some guarded optimism but, if a bright dawn was awaited, there were still to be some dark hours before its breaking. Worldwide slump and massive unemployment were to be the hallmarks of the first years of the next decade and the aircraft industry was to experience some lean years before the quickening drift to another conflict brought expansion toward the massive production of the Second World War.

The Avro 504 had served the Royal Air Force and its predecessors well throughout the 1914-18 War and had continued to do so in the years of peace. From 1927 over 500 of the 504N version with strengthened undercarriage and an Armstrong Siddeley Lynx radial engine were ordered for Service use and construction and repair of these aircraft did much to sustain the Newton Heath factory through the difficult years. Belgium, Brazil, Chile and Greece were amongst the countries purchasing the aircraft and licence construction was undertaken in Siam, Belgium and Denmark. One of the aircraft built in the latter country can still be seen in the Royal Arsenal Museum at Copenhagen.

Although the 504N continued in production, the Royal Air Force was looking ahead toward a replacement and when in 1929 Roy Chadwick designed the first of a series of rugged biplane trainers he drew upon experience with the Fokker-type aircraft built by A.V. Roe & Co Ltd and Armstrong-Whitworth expertise to produce the Avro 621. In 1932 the type was selected as the standard RAF trainer and, as the Tutor, was the first welded steel aircraft to be ordered by the Air Ministry. 795 were to be built at Newton Heath and tested at Woodford whilst production was also undertaken in Denmark and South Africa. As with the many variants of the 504, some Tutors also flew with a variety of civil operators and from the design sprang variants including the Avro 626 general purpose aircraft which, with both Lynx and Cheetah engines, was supplied to over a dozen countries and, as the Prefect, was used by the RAF as a navigation trainer. The 626 included amongst its varied tasks the chasing of hashish runners across the Sinai Desert and eight aircraft were built with increased wingspan as the Type 637 to be armed with two machine guns for the use of the Kwangsi Air Force in South China. Amongst the many well known pilots employed by the Company to test and demonstrate these aircraft and other Avro production of the period were H.A. Brown, S/Ldr. J.L.N. Bennett-Baggs and J.H. (Jimmy) Orrell who was to fly with Imperial Airways and take part in the hazardous wartime service to Sweden before returning to Woodford later becoming Chief Test Pilot.

Whilst final assembly and test flying of Avro aircraft took place at Woodford, many of the aircraft arriving there on their own wheels behind Company lorries, the aerodrome was also being well used by the aircraft of the

A lineup of Avro products at Woodford during the summer of 1930 comprising L-R Avro Six G-AAYR; Avro Five G-AASO; Tutor K1237; Sports Avian G-AAWI of the Lancashire Aero Club and Avian Monoplane G-AAYW.

:Flight International

Lancashire Aero Club. After John Leeming's departure Alan Goodfellow had taken over as Chairman and, drawing upon his wartime piloting experience and the acumen which was building a steadily developing legal practice, he guided the Club into and through the difficult years of depression. The Club now had a new hangar and a comfortable clubhouse, had wiped out a deficit and built up a credit balance at the bank and could boast four Avians and a Cirrus Moth for the use of a membership of over 300. On the 5th July, 1930 the King's Cup air race attracted 88 starters who left Hanworth to fly a virtual circuit of England and the winner was Miss Win Brown in her Avian III, G-EBVZ. Miss Brown, an England international hockey player, was accompanied by another member of the Lancashire Aero Club, Ron Adams, whom she later married. Although she continued to fly for some years afterwards, she had a notable but very different wartime connection with aviation when, after taking her ticket as a trawler skipper, she became Chief Coxwain at the flying boat base at Beaumaris.

The Club continued to train pilots under its successive Chief Flying Intructors and the voluntary part-time instructors, one of whom, Mark Lacayo, was seriously injured on the unlucky 13th April when his Avian crashed near Woodford. Although John Leeming's restless enterprise was now employed elsewhere, there was no lack of initiative in keeping the Club in the forefront of light aviation and in seeking fresh outlets for members' skills and interest.

In 1922 the offer of prizes for motorless flying drew competitors to Itford in Sussex with a motley selection of devices, only thirteen of which flew. F.P. Raynham remained airborne for 1 hour 53 minutes in a Handasyde glider but gliding as a sport did not make any substantial progress in Britain until the formation of the British Gliding Association as a result of a meeting called in 1929 following the German discovery of the technique of making cross-country flights by means of thermal soaring.

The Lancashire Aero Club was not slow to become involved in the mushrooming gliding movement and a Prufling primary glider was acquired from Germany. A gliding section of the Club was formed and B.A.G. Meads was

Miss Winifred Brown in the rear seat of the Avro Avian, G-EBVZ, in which she won the 1930 King's Cup Air Race. *:A. V. Roe & Co Ltd*

appointed Chairman. In later years Basil Meads was to become President of both the Lancashire Aero Club and the Derbyshire and Lancashire Gliding Club and he is today the greatly respected Vice Chairman of the Manchester Branch of the Royal Aeronautical Society. Short flights were made with the Prufling at Woodford and on the 31st July, 1930 the first inter-Club gliding match ever held in Britain took place when a team from the Lancashire Aero Club visited the London Gliding Club at Ivinghoe Beacon. On the 19th/20th October of that year Alan Goodfellow came second in a competition involving six clubs at Ditchling Beacon.

Other gliding clubs were formed in the district including one at Bolton which organised a demonstration at Rivington Pike and the Stockport Gliding Club which purchased the Lancashire Aero Club's Prufling after it had been damaged in a mishap at Sparrowpit. The Manchester Branch of the Royal Aeronautical Society formed a gliding section and members built a Zögling primary glider which made initial hops at Woodford. The Stockport Gliding Club was absorbed and the Zögling and the Prufling were flown for some time from a ridge overlooking Lyme Park at Disley.

With considerable help from Dunlops, Cellon Ltd and A.V. Roe & Co, a B.A.C. VII two-seat training glider was assembled in premises under the present British Broadcasting Corporation studios in Piccadilly and taken to Woodford where it was used extensively for training purposes. The project owed much to Roy Chadwick who was always on hand with practical advice.

In 1935 the Manchester Branch of the Royal Aeronautical Society joined with the Derbyshire Gliding Club to form the Derbyshire and Lancashire Gliding Club and a site at Camphill was leased. A hangar was built and the site is still in use.

The opening of the Airport of Manchester at Wythenshawe had given Manchester a well-deserved lead in municipal proprietorship of a centre for air transport but work to create an aerodrome on the difficult surface of Chat Moss continued and, on the 1st January, 1930, Manchester began a new decade with a new airport at Barton. The official record notes that the first aircraft to land was the Avian, G-AADL of Northern Air Lines piloted by Captain A.N. Kingwill. The first Service aircraft visited on the 6th January when an Avro 504N, J9684 was flown in by Flight Lieutenant Barnes from Sealand and the issue of a permanent public use licence on the 1st April was followed three weeks later by a visit from a Desoutter Monoplane, G-AAPX with Sir Sefton Brancker. The Prime Minister, Ramsey Macdonald, broke his journey to Scotland in the Imperial Airways Argosy, G-EBLF on the 25th May to inspect the aerodrome. His pilot on this occasion was Captain O.P. Jones who, although he had visited Wythenshawe on a number of occasions, had not previously been to Barton which he described — possibly diplomatically — as an admirable landing ground.

The first Royal visitor to Barton was the Duke of Gloucester who arrived in a Royal Air Force Wapiti on the 8th July, ironically to attend an agricultural show at Hough End Fields — the site of the former Alexandra Park aerodrome.

The airport was managed by Northern Air Lines for the Airport Committee whose Chairman, Alderman William Davy (later to be knighted), had long been a supporter of efforts to provide airport facilities for Manchester and the resident manager was William (Bill) Ledlie. Former farm buildings were used as offices and a large hangar flanked by petrol pumps had been erected. The Aerodrome Hotel was ready for opening shortly after the aerodrome and it might well have been named 'The Pitot Head' but this suggestion was turned down by the Airport Committee. Great efforts were made by Northern Air Lines to build up business in the air taxi and aerial advertising fields and the public were frequently encouraged to come to Barton to see flying exhibitions. As early as the 26th January the crowds at the aerodrome were such as to make flying hazardous and when, in July, Barton was a calling place for participants in the King's Cup Race, over 20,000 paid for admission and twice that number were reported as watching from outside the boundary. Only one of the racing aircraft withdrew at Barton, the Avro V, G-AASO flown by F/Lt S.L.G. Pope and amongst the better known pilots to pass through were Lt C John — now Admiral of the Fleet, Sir Caspar John — in an Avian, F/Lt Tommy Rose in a Bluebird and Captain H.H. Balfour in a Spartan Arrow. 61 entrants followed Win Brown back to Hanworth and after they had left, the public at Barton were treated to a display of stunting and crazy flying by Kingwill and N.A.L. pilots Bonar and McKay.

On the 16th June, 1930 Manchester had its first scheduled airline service since the closure of the Daimler route when Captain O.P. Jones brought in the first passengers on the subsidised Croydon-Birmingham-Manchester-Liverpool route. The inaugural aircraft was Argosy, G-AAEJ and the route is usually thought of as an Argosy service, although in fact, Imperial Airways frequently used Handley Page W8F and W10 aircraft as well as the Argosies until flights ceased on the 20th September. Load factors were reasonably good and it was expected that the service would commence again the following Spring but the economic climate was then such as to preclude further development of the route.

Wythenshawe finally closed during 1930 after being used for some time as a secondary landing ground for the training activities of N.A.L. and some thought was given to the establishment of aerodromes at Bolton and White Moss, Middleton but nothing came of these plans, nor was there any development of a suggestion for setting up aerodromes on the roofs of buildings in major cities including Manchester. One field which did see some flying was a part of Blackley Golf Course which was used on numerous occasions by the Gipsy Moth, G-AAKN purchased by Ferranti Ltd in February 1930 for the use of Sir Vincent de Ferranti and for experiments in navigation along a narrow radio beam. The Company made numerous experimental flights both in the United Kingdom and in France and on the 5th June their pioneering work, under the direction of Dr R.C. Galletti, brought the Royal Aircraft Establishment's Fokker FVIIb/3m, J7986 to Manchester. The aircraft was flown from Farnborough to Barton by F/Lt McLaren-Reid and returned to the R.A.E. the following day after flying to the vicinity of Nottingham on a beam transmitted from the Ferranti works. The Fokker, a close relative of the Avro Ten and of the *Southern Cross* of Kingsford-Smith, had been acquired for tests of the Monospar wing and, after its career at Farnborough, ended its days as a coffee bar on Canvey Island.

Aeroplanes were now being seen more frequently over Manchester but the passage of the airship R100 over the city on the 25th May, 1930 was an event sufficiently unusual to provide considerable public interest and was vividly recalled when, a couple of months later, the R100 made a successful flight from Cardington to Montreal. Two months after its return the nation was to be shocked by the fatal crash of the R101 at Beauvais in France with the loss of 48 lives. Amongst the casualties were the Air Minister, Lord Thomson and Sir Sefton Brancker who had done so much to inspire those who had led the efforts to further aviation in and around Manchester during the difficult years of the nineteen twenties. On the 11th October the large bell in the Town Hall clock tower tolled for $1\frac{1}{2}$ hours in tribute to the R101 victims. Not long afterwards the R100 was broken up but Mancunians can still see part of the woodwork from its passenger accommodation incorporated in the furnishing of the Crown and Kettle Hotel on Great Ancoats Street.

From the earliest days at Wythenshawe, Leeming and Holmes had found the financial path of Northern Air Lines Ltd to be both uphill and strewn with pitfalls. As the Company pressed on with its activities at Barton, the joy-riding fleet of Avro 504s ranged widely and gave large numbers of people a first experience of the air. Facilities at Wythenshawe were virtually non-existent and as well as experiencing difficulty in keeping their own aircraft flying N.A.L. found that the outside maintenance and repair work which had provided a source of income for Holmes at Shrewsbury could not be undertaken. A little of this work returned when the Company moved to Barton but capital was hard to come by and, although some customers, notably the newspaper concerns, provided air taxi work, the prevailing trade conditions prevented other well disposed concerns from doing so. Some flying training was carried on and plans were made for a service from Barton to Liverpool, Blackpool and the Isle of Man but Manchester money was in cotton and cotton was having a bad time and little finance could be found. Some comfort was found in the return from occasional displays at Barton and wonders were done with the few serviceable aircraft which could be spared from the joy-riding fleets. Occasionally for these shows aircraft were borrowed — the Avian owned by Peter Eckersley who had succeeded Alan Goodfellow as Chairman of the Lancashire Aero Club was put to good use — and Kingwill and such pilots as George Yuill and Bill Ledlie flew continuously to give the public value. The Manchester Aero Club — successor in name alone to the pre-1914 Club at Trafford Park — was formed in 1932 and had meetings at the Barton Airport Hotel but by mid-1933 the Airport Committee had transferred the management of the aerodrome to Airwork Ltd and shortly afterwards Northern Air Lines Ltd went into liquidation. John Leeming was to write later that the Company had been five years ahead of its time.

The Manchester Aero Club was taken over by the Lancashire Aero Club who placed two aircraft at Barton and continued to provide flying training and a high standard of club facilities at Woodford. No less than for other local organisations, the commercial slump had brought some anxiety but membership and flying activity had been maintained and the Club had continued to play its part in the national structure of club flying. Several newly fledged light aircraft clubs in the North-West — some of them still thriving — had owed their existence to encouragement and assistance from Woodford. Lancashire Aero Club members had taken part with distinction in successive King's Cup races and such events as the 1931 and 1932 Liverpool-Manchester Inter City races. F.R. Hall won the Society of British Aircraft Constructors' Cup at Heston on the 20th June, 1931 in the Club's Avian,G-EBRR and on the 18th June, 1932 the Lancashire Aero Club organised a two-lap race round the Isle of Man, the competitors being shepherded across the Irish Sea from Stanley Park at Blackpool by a Saro Cutty Sark of British Amphibious Air Lines. An interesting aircraft which joined the Club fleet was the Desoutter Monoplane, G-ABRN operated for the British Red Cross Society.

For the man in the Manchester street there were sufficient evidences of financial stringency as many firms went out of business and unemployment increased and the difficulties of local concerns in aviation were, perhaps, not widely known or understood. Aviation continued to come to his notice in various ways: in August 1931, blue air mail post boxes appeared for the first time and from time to time local theatres featured personalities from the world of aviation. Winifred Brown made several stage appearances on one occasion leading a bill including George Formby and Ted Ray, and the physical effort entailed in touring the country for such performances was brought into perspective when Amy Johnson collapsed on stage at Bolton in January 1932. A presentation which provoked wide controversy was that at the Paramount in Manchester in August of the same year when Major Chris. Draper appeared on stage with two German pilots to recount wartime experiences. Draper, a native of Bebington in the Wirral, was also seen over the area towing advertising banners and maintained his publicity image as the Mad Major when, in 1953 at the age of 61, he again hit the headlines with a flight in an Auster under a number of London's bridges.

From time to time the papers carried reports of aviation fatalities. Sadly, the biggest headlines tend to be reserved for the events which do least to enhance the public's confidence in aircraft but crashes are a fact of life and a matter of record and, when an RAF two-seat Siskin from 5 F.T.S. at Sealand crashed at Barton on the 24th May, 1932, there was evidence of the selfless gallantry which is frequently shown on such unhappy occasions. One airman was killed on impact and the aircraft caught fire immediately. The instructor was pulled clear by E.W. (Jock) Bonar of Northern Air Transport and a police constable, but died of his injuries two weeks later. Nine months later Bonar visited Buckingham Palace to be invested with the Empire Gallantry Medal — later exchanged for a George Cross — and P.C. Herbert Hayes received the King's Police Medal. In October 1932 there was another fatal accident, this time at Irlam when the Cirrus Moth, G-EBZU crashed shortly after taking off from Barton but, for the most part, the inevitable incidents involved with flying training in the Manchester area at the time led to nothing more than crumpled aircraft and crumpled pilots, both of which were quickly repaired.

A feature of the aviation scene during the 1930s was the travelling air circus and it is usually the name of Sir Alan Cobham which comes to mind when itinerant air displays are mentioned. Passionately determined to advance the air-mindedness of the nation he toured the country with a collection of aircraft and pilots able to take up passengers and to give a top class display of flying and, whilst existing aerodromes were used from time to time, Cobham also visited many towns to fly from a field rented for the occasion. C.D. Barnard and C.W.A. Scott also toured with travelling pageants of the same kind and it is possible that Barnard was first in the field. He was at Barton on the 29th May, 1931 when his aircraft included the famous Fokker, G-EBTS *The Spider* which, shortly before had had the misfortune to be caught by the tide at Southport, an event unusual and fortunately not causing serious damage. Cobham had made a grand tour of Britain with a joy-riding fleet in 1929 but had excluded the Manchester area on the grounds that he felt he would be preaching to the converted in view of the efforts of Holmes and Leeming. In 1932 he was at Bangor, Rhyl, Crewe, Warrington and Blackpool in one week and his display team visited the local area to perform at both Woodford and Barton in June 1933, following those performances with shows in the last week of July at Harehill Farm, Heywood, Siddon's Farm on Ainsworth Road at Bury, Back of Moss Farm, Unsworth and Manor House Farm, Timperley. Amongst the aircraft used were the Airspeed Ferries, G-ABSI and G-ABSJ and an SE5a registered G-EBVB.

The opening days of 1934 found Manchester with its own aerodrome but with virtually no contact with the air transport services which were beginning to pattern the globe with routes both between the nations and within their frontiers. Britain's major efforts were directed toward linking London with the continent and the outposts of empire and there were few services between the country's towns. On the 23rd January the famous K.L.M. pilot Captain Ivan Smirnoff came to Barton in the Fokker FXII, PH-AII to survey the aerodrome for inclusion in a route to Amsterdam and he was not impressed. The Royal Dutch airliners were to become familiar to many Manchester people and their regularity was observed with some admiration but they were to be seen only on passage over the city on the stage from Liverpool to Hull.

Although the Manchester City Council had been disappointed in their hopes for Barton and immediately commenced to seek a more suitable site for a major airport, the aerodrome was far from the total liability that some have continued to suggest over the years. If not suitable for the most modern airliners even of that period, it was able to handle the internal air services which developed in the latter half of the decade and on the 9th May, 1934 it was used for early experimental air cargo flights by two Airspeed Couriers, G-ACJL and G-ACJT leased by Bouts-Tillotson Transport. Railway Air Services Ltd used the 4-engined de Havilland Express, G-ACVY *Mercury* to inaugurate a Croydon, Castle Bromwich, Barton, Belfast, Renfrew service on the 20th August and thus linked Manchester with a number of major cities. The loss of the K.L.M. service was largely made up when on the 17th June, 1935 Hillman Airways Ltd commenced to fly the route Hull, Manchester, Liverpool, Belfast.

Barton continued to attract large crowds for the annual Empire Air Day displays and Cobham continued to visit the Manchester area with his Display, using amongst other sites Longford Road, Reddish and a field at Alkrington which the Middleton Council had ideas of developing as a permanent aerodrome. His display at Woodford on the 30th May, 1935 was marred by the death of his parachutist Ivor Price whose parachute failed to open.

From time to time Manchester's leading stores have contributed to the enhancement of the average citizens' contact with aviation and on the 4th February, 1935 Lewis's opened an exhibition which included the de Havilland Comet, G-ACSS *Grosvenor House* which had won the 1934 Mildenhall to Melbourne Air Race sponsored by the Australian McRobertson chocolate company. Thirty three years after its triumph the aircraft is today being restored to flying condition by the dedicated engineers of the Shuttleworth Collection at Old Warden.

No summary of flying in the thirties would be complete without mention of the Flying Flea. Over the years much effort has been directed towards the evolution of a small machine which the mythical average man could build, fly and maintain and in 1935 the publication in 'Practical Mechanics' of plans for Henri Mignet's Pou du Ciel led many enthusiasts to believe that they could achieve a short cut to inexpensive aviation. The Daily Express sponsored construction and French-built aircraft embarked on demonstration tours of Britain, one being shown off at Barton on the 10th August, 1935.

Construction of approximately 1,000 Flying Fleas was commenced and the Manchester area had its fair share. At Royton J.O. Bayliffe used successively a Harley Davidson motor cycle engine and the power unit of an Austin 7; The Dale Motor Co of Rochdale used the local golf course to test a machine with a Scott engine and the Oldham Welding and Central Motor Co built the Anzani-engined G-AEJA which attempted to fly at Barton. Also at Barton, Robert Parker tested and crashed a Flea, G-ADDW built in Blackburn and Harry Killick — later to construct his own man-powered aircraft — flew G-AEII on the power of a converted Douglas engine following construction at Ashley Heath. Bands of aspiring aviators formed the Manchester Flying Flea Club and the Manchester Challenge Flying Flea Club but these groups do not appear to have achieved completion of an aircraft.

Many of the aircraft which were commenced were left unfinished as enthusiasm and funds ran out and, following the death of a number of pilots due to inability to pull out of a shallow dive, the wave of optimism which had surged in with the first Fleas receded and died. Today the two Flying Fleas, BAPC 12 and 13, in the care of the locally-based Aeroplane Collection Ltd, remain to recall that episode in the area's aviation history.

Amongst the more substantial organisations to look at the construction of the Flying Flea was the Trafford Park timber company, F. Hills & Sons Ltd. A number of employees completed an aircraft, G-ADOU, and tested it at Barton but their principal achievement was to thoroughly frighten themselves and the firm's management and the project was shelved. However, the company had observed a potential market for a small and inexpensive aircraft and obtained a licence to build the two-seat Czechoslovakian Praga E114 as the Hillson Praga. Powered by a 36 hp Praga engine built by Jowett Cars Ltd, the first aircraft to be tested at Barton, G-AEEU, arrived at the aerodrome on the 24th May, 1936 and a total of 35 aircraft emerged from the Trafford Park factory. Hills' managing director, W.R. Chown, had experience of flying training and, although the Praga did not achieve many sales to private owners, its flyaway price of £385 made it attractive to a number of clubs and Chown set up the Northern Aviation School and Club Ltd at Barton where the little aircraft were a familiar sight until the start of the Second World War. In 1937 Hills built a development designed by Norman Sykes as the Hill on Pennine with a Praga engine and an unorthodox control system but, although a test flight was made at Barton on the 4th February,

Hillson Praga G-AEPJ of the Northern Aviation School & Club Ltd, Barton circa 1938. :Lancashire Aero Club

1938, the aircraft was not successful and passed into obscurity.

With only limited power, both the Flying Flea and the Hillson Praga had several mishaps in the area without particularly serious consequences to their occupants but the growing number of both civil and military flights over the area round Manchester brought occasional more salutary reminders that aviation had yet some way to go before achieving real mastery over the elements. On the 26th October, 1936 Railway Air Services lost the Dragon, G-ADEE when the pilot became lost in fog on a flight from Barton to Speke. The aircraft crashed near Garstang with the loss of two lives and Edgworth Moor near Bolton collected the Puss Moth, G-AAYD, a fortnight later. Handley Page Heyfords of 102 Squadron, RAF met fog and icing on a flight from Aldergrove to Finningley on the 12th December, 1936 and three came down in the Manchester area. K6898 landed in a field beside Jackson's Edge Road at Disley and collided with a hedge and telegraph pole without injuring its crew. A second machine, K4868, landed safely in the same field but the pilot of another aircraft, K4874, after circling Oldham for some time, found no potential haven and the crew abandoned ship, the Heyford crashing at Dingle Farm, Sholver. One of the crew landed on the roof of the Crabtree Mill and another broke a leg after landing on the roof of a cottage and continuing his descent without benefit of parachute. Another Heyford, K6875, this time from 166 Squadron, crashed at Broadleebank Tor overlooking Edale during a night navigational exercise on the 22nd July, 1937 with the loss of the crew of six.

There were, however, happier occasions when military aircraft visited Manchester. Aircraft from Sealand and other stations continued to visit Barton on training and communications flights and on the 7th April, 1934 nine Hawker Audaxes of 26 Squadron arrived from Catterick for the Royal Air Force's first official Sqadron visit to the City. The Squadron flew in formation over the Manchester area on the following day and lunched with the Lord Mayor at the Airport Hotel.

When Captain Smirnoff made his crushing pronouncement on the adequacy of Barton for international services, Manchester's Airport Committee and the relevant City officials did not waste time. Sites at Audenshaw, Bury and Mobberley were inspected and some thought was given to the use of the existing aerodrome at Woodford but on the 25th July, 1934 the City Council approved by one vote the construction of an airport at Ringway on the southern boundary of the City. Alderman Tom Regan of Gorton, who was to be a member of the Airport Committee for over forty years, always maintained that it was his vote which tipped the scale!

Opposition from residential groups anxious that any disturbance should be strictly for someone else continued but a Public Inquiry approved the purchase of 664 acres of land by the Corporation. Air Ministry approval followed and by the 28th November 1935 the Lord Mayor, Alderman T.S. Williams, was able to cut the first sod to start the work of levelling and draining the area.

Whilst Britain drew away from the dark days of commercial depression, new forces were manifesting themselves in the countries of Europe and amidst the pacifism and complacency of large sections of the government and populace there were some far-sighted realists who saw early that the war to end wars had failed to overcome the aggressive ambitions of a Germany now under the leadership of Adolf Hitler and the Nazi party. Under the lash of Winston Churchill's powerful oratory, Parliament approved an expansion scheme for the Royal Air Force and the aircraft industry had some reason to believe that the lean years had ended.

Production of various aircraft for the civil market and military biplanes for the RAF and foreign air forces had brought A.V. Roe & Co Ltd through the period of industrial depression and in 1933 the Company had acquired a licence to produce the C30A autogiro under the designation Avro 671 Rota. This was the latest of a series of autogiros designed by the Spaniard Juan de la Cierva and built by the Company. One of the earlier machines, a C8L, G-EBYY, was the first rotary winged aircraft to cross the English Channel and can still be seen in the Musée de l'Air near Paris. Examples of the Rota have also survived and can be seen in the Science Museum in London, the Shuttleworth Collection at Old Warden and the Skyfame Museum at Staverton. Over 60 examples of the Rota were built at Newton Heath and several were used by the RAF for wartime radar calibration duties.

The first benefits of the Royal Air Force's expansion programme reached Newton Heath in the form of orders to produce 287 Hawker Audax biplanes and a side effect of these orders was the development of an improved version which replaced the Rolls-Royce Kestrel of the standard Audax with a 750 hp Armstrong Siddeley Panther for the Egyptian Air Force. Despite the value of the Audax orders, it was clear that the era of the biplane was ending. Far-sighted planning at the Air Ministry had led to the ordering of a new generation of fast eight-gun fighters and the Bristol Blenheim which was to supersede the biplane light bombers was being put into production. A new government-sponsored factory at Chadderton was to be opened by A.V. Roe & Co Ltd in 1938 for the production of Blenheims but at Newton Heath the major contribution toward preparation for war was the building up of production facilities for the Anson coastal reconnaissance bomber.

In 1933 Roy Chadwick had designed the Avro 652, a low wing monoplane with two Armstrong Siddeley Cheetah engines, and in the following spring Imperial Airways ordered two for use on flights to European capitals with His Majesty's mails. Delivery was effected in 1935 and the two aircraft were given the names *Avalon* and *Avatar* respectively. The latter name was quickly abbreviated

Avro Anson's from the first production batch in service with 48 Squadron RAF.

to *Ava* when it was found to mean something quite disgusting in a Central European language! The Avro 652 was notable as the first Avro aircraft and the first for Imperial Airways to have a retractable undercarriage.

Within two weeks of the delivery of the civil machines the prototype Avro 652A Anson coastal reconnaissance bomber K4771 made its first flight at Woodford in the hands of S.A. (Bill) Thorn. After evaluation at Martlesham Heath the type was accepted and such was the urgency with which production was tackled that almost 1,000 aircraft had been supplied by September 1939. Many variants of the Anson were to be evolved and when the last aircraft of the type was delivered in 1952, nearly 11,000 had been built.

The Anson had a wingspan of 56′6″ and an all-up weight of 8000 lb. in its initial form but the other Avro aircraft ordered in the mid-thirties was an altogether heftier proposition. Designed by Chadwick to the Air Ministry Specification P13/36, the Manchester was a mid-wing bomber powered by two Rolls-Royce X-type engines. A crew of seven was carried and the all-up weight was 56,000 lb. Stressed-skin construction was used and at the time when the order was given a greater advance on the biplane bombers of the day could hardly have been imagined. Initially fitted with twin fins and rudders, the Manchester had a third fin added but retrospective modification later replaced this tail assembly with a developed twin fin arrangement. The prototype was constructed at Newton Heath and made its first flight at Ringway on the 25th July, 1939 in the hands of Captain H.A. Brown and production of a batch of 200 aircraft commenced at about the same time.

The timely expansion of the Royal Air Force, inadequate though it may have been, did much to provide employment on the North side of the city and at Woodford where the aircraft from Newton Heath and later Chadderton were taken for final assembly and flight testing. On the South side of the city the mid-thirties saw the arrival of a new manufacturer in the area when, on the 30th November, 1934, the Manchester Evening News announced that the Fairey Aviation Co Ltd had acquired the Heaton Chapel factory which had in wartime been the National Aircraft Factory No 2 and, latterly, the car body works of Willys Overland Crossley. The Director in charge was Major Thomas Barlow who had served in the Royal Naval Air Service before becoming Chief Technical Officer at the Martlesham Heath experimental centre and he set about preparing the factory to deal with large orders for the Merlin-engined Fairey Battle. Before the Battle production got under way, however, 14Hendon bombers were built and taken to Barton for testing by Flight Lieutenant Duncan Menzies. The Hendon was a cumbersome twin engined aircraft with two Rolls-Royce Kestrels and rapid development of more advanced bombers led to the cancellation of orders for further aircraft. The Hendons produced all went to 38 Squadron at Mildenhall, alhough some were borrowed briefly by 115 Squadron at Marham.

The first Battle from Heaton Chapel, K7558, was test-flown at Barton by Flight Lieutenant Menzies on the 14th April, 1937 and deliveries to the RAF's 63 Squadron at Upwood commenced on the 20th May. Twenty-one aircraft were flown at Barton before testing was transferred to the first available section of the new aerodrome at Ringway. The lease of land there had been agreed on the 11th March, 1936 and by the 2nd June, 1937 the hangar which is today used for work on Fairey Britten-Norman Islanders was opened by the then Lord Mayor, Alderman Joe Toole. Even before the testing of Battles was commenced at Ringway, the Hornet Moth, G-ADND was making frequent visits usually in the hands of Duncan Menzies.

Whilst the testing of Battles went ahead, work continued on the construction of the terminal building and main hangar at Ringway. The Council were encouraged by the publication of the Maybury Report on the 25th January, 1937

The first production Fairey Hendon, K5085 running up its engines at Barton possibly just before its first flight. *:R. Cheeseman*

A Heaton Chapel built Fairey Battle of 63 Squadron RAF in the camouflage and markings applied during the Munich crisis. :Flight International

when the establishment of a focal airport to radiate routes to many other towns was advocated and there was a firm belief that Ringway would become the designated airport. Sam Hill was appointed Airport Director and on the 25th June, 1938 Manchester's new airport was officially opened by the Air Minister, Sir Kingsley Wood. A service to Amsterdam was inaugurated by the K.L.M. Douglas DC2, PH-AKP and a display which included both civil and RAF participation gave many members of the public an encouraging introduction to the upsurging aviation scene. Scheduled services were operated throughout the summer of 1938 and increasing number of small civil aircraft visited the airport. The Munich crisis of September 1938 caused a temporary cessation of the K.L.M. service but this was resumed in the following year along with additional internal services using mainly the de Havilland Dragon Rapide and the four-engined de Havilland DH86 Express.

On the 20th May, 1939 the Air League promoted an Empire Air Day Display at Ringway with large-scale RAF participation. A large crowd attended and were able to see a Supermarine Spitfire, formation flying by Whitleys and Tiger Moths and such relatively elderly types as a Heyford, an Overstrand and a Seal. Locally built aircraft including a Battle and an Anson took part in the display which was marred by the crashing of a Lysander flown by Pilot Officer H.G. Malcolm. Although seriously injured, Malcolm recovered to win a posthumous Victoria Cross as a Wing Commander in North Africa.

The Manchester people who flocked to Ringway for the display had their first sight of aircraft belonging to a unit which they were to know with pride as their own. The first Auxiliary Air Force Squadron had been formed in 1925 and as early as the 25th September, 1930 the Evening Chronicle had reported pressure for the formation of an auxiliary unit to be based in Manchester. It was not until

1939 that the East Lancashire Territorial Army and Air Force Association was to be enabled to play its part in the formation of a squadron which was to be known throughout a distinguished wartime career and an all too brief period of peace time activity as No. 613 (City of Manchester) Squadron. On the 1st February, 1939 the Squadron, based at Ringway, took its place in No. 22 (Army Co-operation) Group of Fighter Command and on the 11th May Alderman E.J. Hart, the Lord Mayor broke a bottle of champagne over the undercarriage of a Hawker Hind trainer, K5473, to mark the City's support for its own unit. The Squadron badge incorporated the fleur de lys of the Manchester Regiment and the motto 'Semper Parati' was adopted. After training had got under way at Ringway, 613 Squadron went to Hawkinge in Kent for its first summer camp.

Barton continued to receive a varied selection of aircraft as private flying boomed and Service visits increased. A glance at the tower log reveals visits by such varied types as an RAF Valencia, K2344, the Lockheed 10, OK-CTA, of the Bata Shoe Company and Sir Kenneth Crossley's Hornet Moth, G-ADKC, the registration of which was a forerunner of many present-day personalised allocations. On the 1st October, 1937 Airwork Ltd brought six Gipsy Moths to be used by No 17 Elementary and Reserve Flying Training School of the Royal Air Force, the unit's first headquarters being at Sunlight House in Quay Street, Manchester. An Air Ministry control centre was established at the aerodrome on the 21st October, 1937 and when the Civil Air Guard was set up in July 1938 to train men and women at cheap subsidised rates the Northern Aviation School and Club soon adopted the scheme. The Lancashire Aero Club also trained pilots of the CAG at Woodford. Although Barton was never regarded as a big aerodrome, it successfully received and dispatched quite large aircraft including a Wellington, L4284 of No. 148 Squadron from Stradishall on the 11th April, 1939 and a Bombay, L5813 which arrived from Odiham on the 26th July, 1939 and made no fewer than ten local flights before departing.

Aircraft were becoming a far more common sight over the Manchester of the late nineteen-thirties and some of them made quite deliberate efforts to be seen by as many people as possible. Banner towing biplanes proclaimed the virtues of various products and a lot of older residents will recall seeing the word 'Persil' written laboriously in whiter than white smoke in a cloudless blue sky. Even in those days there were also people who foresaw travel beyond the earth's atmosphere and in 1936 the Manchester Interplanetary Society was formed under its youthful President, Eric Burgess who, with the Society's Secretary, Malcolm Wade, organised what may well have been the world's first model rocket competition at Clayton Vale on the 27th March, 1937. The competition ended disastrously with a spectator being injured when one of the rockets exploded but forecasts made by the Society's members forty years ago have been proved by later events to be well-founded in many respects. Experiments continued in the next two years but emphasis was placed more on theoretical exploration of the means of developing rocket propulsion.

The Munich Agreement deferred the onset of war but few had faith in the Prime Minister's forecast of 'Peace in Our Time'. Throughout the summer of 1939 civil aviation continued to develop but the expansion of warplane production and the training of pilots continued apace and events in Europe cast a deepening shadow. On the 1st September German forces invaded Poland and British forces were mobilised.

Aviation had awakened. Its day was to be opened with a burgeoning force which would leave an indelible mark on the Manchester area and the World.

The Second World War

When Neville Chamberlain sat gravely before a microphone at 11.15 on the 3rd September, 1939 to announce to the nation and the world that a state of war existed between Great Britain and Germany, he might well have also announced that from that moment the relationship between the people of Manchester area and the various manifestations of flight would be totally and irrevocably changed.

If the First World War had brought massive advances in aviation, those advances were to be dwarfed by the progress in technology, production and utility which the 1939-45 conflict was to bring. Whereas few people in 1939 had more than passing acquaintance with the limited numbers of flying machines then to be seen in the area, most were to be daily and tangibly aware of the great fleets of aircraft built up during the war period by the Royal Air Force, its allies and its enemies. When the war ended the relationship had changed and aviation had become a decisive feature of war and a factor of the peace-time scene which, if it was not a part of the daily life of the average citizen, was constantly brought before him in his newspapers, on his radio set and, later, on the television screen. Throughout the war the Manchester skies saw aircraft in large numbers and variety. Many Manchester peope were employed in building them and many more went away to fly them or to maintain them. All, at home or away in the Services, could hardly fail to be constantly aware that aircraft had become an almost ever present source of both possible harm and potential protection.

It is impossible within the space and scope of this booklet to do more than sketch in flimsy basic outline the many aspects of Manchester's involvement with war-time aviation. It may be that the following pages will remind older readers of events in which they took active or vicarious part or will serve to stimulate in later generations a desire to look more deeply into records of those days.

The most immediate visual intrusion of aerial matters upon the local population at the outbreak of war came in the form of preparations for protection against hostile action by the enemy. As the children were evacuated to allegedly safer areas and various forms of air raid shelter were hastily provided, barrage balloons appeared in the sky over the city and the surrounding districts. No. 10 Balloon Centre, Royal Air Force, had been set up before the war commenced and Nos. 924, 925, 926 and 931 Squadrons controlled balloons from a large number of parks and open spaces with small units of RAF personnel living in huts beside their charges. From time to time high winds caused balloons to break from their moorings and career away to provide target practice for the Yorkshire-based fighter squadrons. As manpower needs increased, many of the balloons were handled by personnel of the Women's Auxiliary Air Force (WAAF).

Manchester was a centre for production of barrage balloons, over 30,000 emerging from the Dunlop factory set up in former gasworks premises at Gaythorn and others being produced in the City Hall off Deansgate. Four thousand balloons were shipped to America and used for the protection of the Panama Canal Zone and United States Pacific Coast.

The threat — and subsequent fact — of air raids also brought anti-aircraft guns and later various odd rocket devices to sites scattered about the area and, whilst aircraft recognition became a feature of the lives of many factory workers doubling as raid spotters, personnel of the Royal Observer Corps manned full time the posts which had been hastily built up, initially of sandbags and later of more permanent materials. No. 7 Group Headquarters and Operations Rooms was at the Manchester GPO in Spring Gardens from 1937 until 1942 when an old house in Grange Avenue, Levenshulme was taken over and wartime command was exercised first by Lieutenant Colonel E.G. Tuite-Dalton MC and from 1940 by Observer Commander A.R.C. Huntingdon. Throughout the war local posts continued to report not only the assaults of the enemy but also vast numbers of friendly aircraft movements, frequently enabling lost aircraft to be safely recovered or diverted from high ground. In 1944 the local posts reported the only flying-bomb raid against the area and the alertness of the Corps' personnel was ably demonstrated when, on the 29th December, 1944, a number of South Manchester posts reported a V2 rocket which fell on London, almost 200 miles away.

To cope with the intake of personnel for the rapidly growing Royal Air Force, a number of recruiting and medical centres were opened in whatever premises might be available and the author recalls from personal experience the draughty austerity of the requisitioned corridors of the former Manchester Girls' High School in Dover Street, Chorlton-on-Medlock where many local men were given their first contact with Service medical examination procedures. At Heaton Park where at the end of the First World War the RAF had had a Dispersal Centre, hutments appeared again and the park has its place in flying history as the centre through which most of the aircrew personnel destined for overseas training or operations passed on their way to the West coast ports. Further afield a large camp was set up at Padgate, near Warrington to receive the vast influx of recruits for initial training and other similarly raw airmen pounded the promenade of Blackpool. Other recruits were absorbed at the new camp at Wilmslow where in 1940 No 6 Recruit Centre was set up. Before the Station closed over 200,000 airmen were to pass through and one of these achieved the historic Service number, 1234567, in 1941. In addition to the many airmen, thousands of members of the WAAF were also trained at Wilmslow. Large RAF stores establishments sprang up at Handforth and Heywood.

Manchester's own 613 Squadron was embodied into the Royal Air Force on the 23rd August, 1939 and strengthened by the addition of regular personnel to its local volunteers. After a spell at Odiham the Squadron moved to Hawkinge, scene of its first and only pre-war training camp, and as the British Expeditionary Force drew back to the Channel ports, earned undying glory as it dive-bombed German positions near Calais in obsolete Hawker Hector biplanes. The few ill-armed Westland Lysanders which had been received were used to drop supplies to British troops in the same area with the Hectors flying as escorts. Curtiss Tomahawks and North American Mustangs were subsequently used and it was with the latter that 613 Squadron returned briefly to Ringway in March 1943 to fly low over the City in support of the 'Wings for Victory' campaign. One of the Mustangs was placed on display in Piccadilly. De Haviland Mosquitos replaced the Mustangs in late 1943 and many attacks were made on targets in France, Holland and Belgium, the most famous of these being the highly successful raid on the Gestapo Headquarters at the Hague led by Wing Commander R.N. Bateson, DFC. At this time 613 Squadron was under the command of Air Chief Marshal Sir Trafford Leigh-Mallory who was later killed whilst flying to take up an appointment in the Far East. He is commemorated in a stained glass window in the Parish Church near his birthplace at Mobberley. The City of Manchester Squadron crossed to

France in November 1944 and continued to harass the enemy by day and night, some of its sorties extending to Berlin. The Squadron's last operation took place on the night of the 26th/27th April, 1945, when the Mosquitos attacked railway targets in the area around Fuhlsbuttel.

If 613 Squadron was truly Manchester's own, it should not be forgotten that another Royal Air Force unit has carried the City's name in a less spectacular manner for many years. In the early years of the war the RAF set out to capture for itself a sizeable proportion of those young men who had undertaken courses at the country's universities and the Manchester University Air Squadron was one of a number of units set up. Formed on the 7th February, 1941 with headquarters in the Old Christie Library, the Squadron had as its first Commanding Officer Squadron Leader Ernest Rhodes who had been the first to command 613 Squadron. Flying experience was limited as only one aircraft, de Havilland Tiger Moth,W7952, was acquired but, as well as ground instruction, air experience in a variety of other Ringway-based aircraft was available. Many pilots who took a worthy part in the latter years of the war had their first training with the Unit.

The Manchester University Air Squadron was formed to attract to the Royal Air Force young men whose academic prowess made them suitable for aircrew service but there were very many more who, if less fortunate scholastically or financially, were keen to prepare for a place in the Service either in the air or on the ground. As early as 1938 Roy Dobson had been instrumental in the formation of the first local unit of the Air Defence Cadet Corps and when on the 5th February, 1941, the Air Training Corps was formed, many Squadrons, some of them based on ADCC units, were inaugurated in and around Manchester. As the war progressed, thousands of cadets progressed through the Squadrons to service in the RAF and as early as August 1943 one of them, Flight Sergeant A.L. Aaron who had joined up from No. 319 (Broughton) Squadron was awarded a posthumous Victoria Cross for gallantry when flying as captain of a Stirling on a raid on Turin. In addition to those who joined the ATC as cadets, young men on deferred service whilst waiting for places at Service flying training establishments took initial instruction with the Corps.

Every effort was made to give cadets air experience when possible and to provide periods of training at RAF Stations. Obsolete aircraft were allocated to ATC Squadrons for training purposes, local examples included the two experimental Gloster F5/44 fighters which arrived at Oldham, an Avro Cadet at Wigan and a Spartan Cruiser at Cheadle. The Gloster Gauntlet, K5289 was allocated to No. 1855 Squadron at Royton, a unit which has continued to distinguish itself to the present day and a Westland Wallace, K6054 ended its last flight at Barton to be used by a local squadron.

Another book would be required to detail the hazards brought to Manchester and the surrounding districts by aviation in the service of His Majesty's enemies and, whilst the North West did not suffer the intensity of bombing which beset some of the cities nearer to the continent of Europe, the drone of German aircraft and the whistle of bombs became all too familiar. The Manchester 'Blitz' on the nights of 22nd and 23rd December, 1940 brought vast damage and many casualties and there were many other nights when bombs fell on both homes and factories but, in common with their counterparts in other areas, Manchester people carried on with their efforts to support and sustain the armed forces. The defences were not entirely without successes and residents of Hazel Grove still recall the night of 8th May, 1941 when a German aircraft crashed on Torkington Golf Course after part of the crew had succeeded in escaping by parachute.

The progress of the war reduced the attacks on the North West and carried a great revenge to German cities as Manchester hit back through its factories, both Lancasters and Halifaxes built and tested locally repaying in overwhelming measure the Nazi attacks. The final assault directed at Manchester came on the morning of 24th December, 1944 when a fleet of Heinkel He111 bombers flew up the East Coast to the vicinity of Hull and released approximately 40 V1 flying bombs. These pilotless aircraft crashed to earth at places as widely separated as Durham and Shropshire but a few reached the Manchester area and casualties were caused at Abbeyhills Road, Oldham, Garners Lane, Stockport, Worsley and Tottington. Other flying bombs fell at Radcliffe, Hyde and Didsbury, the last mentioned being one of a few which, before exploding, released propaganda leaflets purporting to include letters from British prisoners of war in Germany.

For every German aircraft which flew over Manchester there were very many more Royal Air Force machines as the war effort built up and activity at the local aerodromes grew beyond the wildest expectations of a few years before.

Woodford installations grew with the massive upsurge of Avro production and the aerodrome which had been opened to receive small biplanes from Newton Heath expanded to see large numbers of potent warplanes leave its runways for service with the Royal Air Force. The Lancashire Aero Club ceased operations for the duration and many of its members joined the flying Services or Air Transport Auxiliary. The Club's aircraft disappeared as they were impressed into the Royal Air Force, although two Avro Cadets spent the wartime years in storage. Cranage, near Middlewich contributed many hours of flying to the training activities of the Service and, as the United States began to send aircraft in large numbers for its own and also the British forces, Burtonwood near Warrington launched an increasing number of new types to share the skies over Manchester. Nearby at Stretton, the Fleet Air Arm contributed its share, many of its aircraft having been fabricated at the Fairey factories at Heaton Chapel and Ringway.

The outbreak of war saw a complete upheaval of the activities at Barton. Within a short time the firms which had used the aerodrome in peace-time were required to leave as the field was requisitioned for National Air Communications and Air Taxis Ltd, Personal Airways Ltd and Wrightways Ltd moved in. Civil flying without special permit was prohibited and the aircraft which the new residents flew on communications work soon acquired camouflage and Service serial numbers. The National Air Guard ceased to operate and the Airwork operated aircraft of No. 17 E. & R.F.T.S. were flown away.

Both Personal Airways and Wrightways failed to survive the war years, although for some time the latter carried out aircraft and engine repair work at Barton. Air Taxis Ltd took on the repair of Avro Ansons for the RAF and expanded their work force as they filled the main hangar with damaged aircraft brought in by road and other Ansons flown in for overhaul. The Airport Hotel was commandeered in 1940 and remained in Service hands until after the cessation of hostilities.

Although it was not felt necessary to lay down runways at Barton, a surprising selection of heavy and potent aircraft visited the aerodrome from time to time. Wellington, Whitley, Dakota and Mosquito aircraft all successfully landed and left in safety and on the 25th April, 1944 a Flight Sergeant Crawford brought from Waterbeach a Lancaster, ND920 which circled the field before landing without apparent difficulty. The crew had lunch in the Air Taxis canteen whilst various suggestions for dismantling the aircraft were made as it was thought that it had little chance of taking off in the available distance. The pilot insisted that he could leave safely and the Lancaster was dragged back to the edge of the aerodrome, run up to full power and released to take off with ease and return to Waterbeach.

Various types of fighter aircraft used Barton, notable visitors including Squadron Leader Bader (later Group Captain Sir Douglas Bader) who brought his Spitfire, P4431 of 222 Squadron from Kirton-in-Lindsey on the 13th May, 1940 and Pilot Officer Peter Brothers (later to reach air rank in the post-war Royal Air Force) who flew in from time to time. A Lockheed Lightning visited in 1942 and Martlets and a Buffalo also added variety to the Hurricanes and Corsairs which became common as the war went on.

Civil air transport returned to Barton in 1940 and, until the service reverted to the use of Speke in 1942, West Coast Air Services Ltd and Aer Lingus operated DH86B Express air liners to Dublin, the latter Company occasionally using Douglas DC3s on the route. The service was not without its hazards and on the 18th October, 1941 the DH86B, G-AENR was damaged by gunfire from a 'friendly' merchant ship.

The Ministry of Aircraft Production took over the control of Barton from National Air Communications and throughout the war years the supervision of activities at the aerodrome was in the hands of Mr J.B. Lee who had been employed by Manchester Corporation before the war and returned to their employ at Barton in 1946. Four new hangars were built during the war and one of these was taken over by Hills for the erection of Percival Proctors.

Another hangar was occupied by David Rosenfield Ltd, well-known Manchester motor dealers and a subsidiary of Henlys Ltd. When the war broke out the sale of motor cars for civilian use was heavily curtailed and the Company undertook sub-contract work on Fairey Battles. The Manchester Ice Palace was taken over and Hurricanes were brought to the building for repair and modification. Conversion of Hurricanes for catapulting from Merchant Navy ships was also carried out at Cheetham. As the work expanded, the hangar at Barton was used for assembly and air testing and large numbers of Fulmars and Corsairs followed the Hurricanes.

Ringway was not requisitioned and remained under the management of the City Council throughout the war. The use of the aerodrome for the training of Britain's airborne forces is well known and commemorated by a colourful glass memorial in the main concourse of the present airport buildings. It is also fairly generally known that many aircraft produced in local factories were tested there but other activities at the airport are, perhaps, not as well recorded.

At the start of the war considerable efforts were made to camouflage the buildings and to protect some of the installations with sandbags. The concrete circle and the name 'Manchester' in the centre of the landing area were obliterated to avoid providing a tempting target for German bombers. In fact, the only bombs which fell at Ringway came closer to removing the concrete circle than the buildings.

Under the National Air Communications scheme Rollason Aircraft and Engines Ltd moved into the main hangar adjacent to the terminal to repair the many civilian aircraft which had been impressed into the Royal Air Force and the Company also carried out work on similar aircraft based at Ringway for other purposes. In January 1940 they moved into another hangar, handing over their initial accommodation to A.V. Roe & Co Ltd to supplement the overburdened facilities at Woodford.

Several operational Squadrons spent short periods at Ringway, although none was based for any length of time. Immediately after war was declared, eleven Hampdens of No. 83 Squadron arrived for a brief stay but in the main the visitors were attached for defensive purposes and they included Hurricanes of 253 Squadron based at Kirton-in-Lindsey, Spitfires of 64 Squadron at Leconfield and Defiants of 264 Squadron which as well as enjoying occasional detachments to Ringway, used the airfield as a refuelling base for patrols over the sea approaches to Liverpool from their base in Lincolnshire. Additional fighter cover for the Manchester area came from Squadrons based at Cranage and after 1940 Ringway was well occupied by other units. One of these was No. 6 Anti-Aircraft Co-operation Unit which used a miscellaneous collection of former civil aircraft including de Havilland Dragonflies, General Aircraft Monospars and Short Scions. One of the Scions, AV981, crashed at Lode Hill on a landing approach on the 1st November, 1940 with the loss of the crew of two and the unit lost the de Havilland Leopard Moth,W5783 on the 19th April, 1941 when it struck a brazier on landing and was destroyed by fire. A notable aircraft with the Station Flight during this period was Jean Batten's Percival Gull, G-ADPR (AX866) which can now be seen in the Shuttleworth Collection at Old Warden.

The Air Transport Auxiliary established No. 14 Ferry Pilots Pool at Ringway and, as well as taking away the hundreds of aircraft produced in Manchester factories to the Royal Air Force, brought many types of aircraft to Ringway to transit.

Although there were many facets to the airport's contribution to the nation's war effort, there can be no doubt that it is for its part in the creation of Britian's airborne forces that Ringway will be best remembered. On the 22nd June, 1940 Winston Churchill gave instructions for the creation of a corps of parachute troops and within a month the first live jumps were made. The Royal Air Force took responsibility for the training of the soldiers and by the time the war ended over 400,000 descents had been made. Formed initially as the Central Landing Establishment, No. 1 Parachute Training School commenced with little equipment and few aircraft, parachute training being carried out from Armstrong Whitworth Whitleys which had been grudgingly conceded by Bomber Command and which, although they were universally regarded as eminently unsuitable, were used in growing numbers until they were eventually replaced by Douglas Dakotas. The first jumps were made from platforms built to replace the rear turrets of the Whitleys but this practice was soon discarded in favour of static-line exits from a hole in the floor of the fuselage. Tatton Park was taken over as the principal dropping zone and the trainees made their preliminary drops from baskets hoisted aloft beneath balloons. As well as hardened troops from virtually every country on the Allied side, No. 1 P.T.S. also trained many agents to be dropped behind enemy lines and these included a number of women.

In addition to the massive efforts devoted to the training of parachutists, Ringway also saw the first experimental steps which ultimately led to the creation of the great fleets of military gliders which carried airborne troops into action in the latter years of the war and, alongside the Whitleys, a small fleet of pre-war single — and two seat — gliders was gathered together. The Airborne Forces Experimental Establishment was formed at Ringway and subsequently moved to Sherburn-in-Elmet in July 1942. Work continued there on the Hafner Rotachute which was designed to provide an alternative to the parachute and consisted of a two-bladed rotor surmounting a seat in front of a small tapered fuselage. Models were dropped from a Boulton Paul Overstrand over Tatton Park and tests were

American volunteer pilot P/O H.H. Strickland from Detroit in one of the Manchester Presentation Spitfires (believed to be Mk. Vb AA881) which served with 71 (Eagle) Squadron. :Imperial War Museum CH4266

made at Ringway with a piloted Rotachute towed behind a Tiger Moth and a Tutor.

No. 1 Parachute Training School remained at Ringway until after the end of the war as the airport's installations expanded and large buildings were erected for both military use and the assembly and testing of aircraft. Although there was maximum effort toward the prosecution of the war, the Airport Committee found time to look ahead optimistically to better times and as early as September 1943 the magazine 'Flight' reported ambitious plans for a new terminal to cater for flights 'to the Americas and Asia.'

With so many aircraft operating in the area and from the many operational aerodromes to the east of the Pennines it was not surprising that some failed to return to their bases safely. The incidence of crashes, whilst not great in relation to the hours flown and the hazards created by blackout conditions and the need to fly often in unfavourable conditions, rose sharply in comparison to the peace-time years. High ground in the Peak District and Pennines brought untimely ends to aircraft of a wide range of types, both on operational and training flights and there were other incidents which drew sharp attention to the dangers of wartime flying.

On the 3rd August, 1943 a Wellington, HE703, of No 27 O.T.U. crashed on allotments between the Bridgewater Canal and Walton Road, Sale, two of the crew being killed and a Lancaster, PB304, of No 106 Squadron came down close to Whit Lane, Pendleton on the morning of the 30th July, 1944 with the loss of the crew and a number of civilian casualties. Not long afterwards, on the 4th

October, 1944, a Halifax ploughed into the churchyard of St Andrew's, Blackley, some of the crew escaping by parachute.

With many aircraft being lost on operational flights, a further toll being taken by the elements and an ever-increasing demand for increased efforts to produce more and more aircraft for the RAF and Fleet Air Arm, the population were constantly exhorted to provide scrap metal for melting down and money to pay for the growing production. As the growing ferocity of German U-Boat attacks took toll of ships bringing vital metals, park railings disappeared and housewives were pressed to give up their aluminium pans to be reconstituted into fighters. The emotive attraction of buying a fighter for £5,000 led to the creation of Spitfire Funds in every area large and small. Manchester supported its savings campaign with, amongst many other schemes, the sale of special Spitfire tickets on the trams and buses and Stockport also issued similar tickets. Collection of £5,000 for the funds was rewarded by the application of the district's name to a suitable aircraft and a selection of locally adopted Spitfires included P8379 *Oldham and Lees*, P8526 *Bredbury and Romiley*, R7215 *Stockport* and many more. Manchester itself provided funds for a number of Spitfires, including AA873 *Manchester Air Cadet* and AA880 *Manchester Smithfield Market*.

To encourage the public to subscribe to the various funds, aircraft were displayed in Manchester and surrounding towns and, particularly at the time of special efforts to encourage war savings, large exhibitions and processions were arranged. At various times captured German aircraft were exhibited at Cheadle, Wilmslow, Ashton, Bolton and Prestwich and the 'Wings for Victory' campaign of 1943 brought Marshal of the Royal Air Force Lord Trenchard to Manchester to open a large exhibition at Piccadilly where a Lancaster, ED749 and a Mustang of No 613 (City of Manchester) Squadron were on show. Other Lancasters flew over the site and 613 Squadron put up a full squadron fly-past before being entertained to dinner at the Town Hall. A Fairey-built Halifax and Fulmar were placed on display in Stockport and another Fulmar supported Ashton-under-Lyne's efforts. Not only 'Wings for Victory' attracted support but 'Warships Weeks' and 'Salute the Soldier Weeks' also brought aircraft to be displayed alongside the weaponry of the other Services, typical being the Spitfire which graced Piccadilly in April 1944. In all these campaigns the major stores played a considerable part and Lewis's, Kendal's and Paulden's all had exhibitions during the war years.

Manchester's total involvement in the prosecution of the war took many forms but it could be argued with some force that the area's major contribution lay in the production of vast quantities of material for the fighting Services and in this the aircraft factories played an outstanding part. It is probably true to say that of any collection of people asked to name Manchester's greatest contribution to the winning of the war, the majority would name the Lancaster bomber but, dominant though that worthy aircraft may have been, it was far from being the only effective flying weapon to emerge from local factories or, indeed, from the Avro works. In 1939 the Anson, one of the best loved and longest serving aircraft ever to be used by the RAF, was already in service in both the coastal reconnaissance and training roles and as the war progressed production at Newton Heath and Chadderton was supplemented by manufacture in Canada as the growing need for aircrews brought expansion of flying training establishments in that country and throughout the British Commonwealth. Until superceded by more powerful aircraft, Ansons patrolled the coasts, and despite their limited armament — occasionally supplemented by enterprising units — succeeded in taking their own toll of enemy aircraft. Air-sea rescue operations followed and many Ansons gave docile and reliable service as light transports, not least to the widely travelled crews of the Air Transport Auxiliary. Blenheims continued to flow from Newton Heath to Woodford and on to the RAF. On the 16th October, 1940 one of them, Z5964, was wrecked on a delivery flight, the pilot, who was seriously injured, being Robert Smith Barry who had so strongly advocated the use of the Avro 504 for pilot training in the First World War. Blenheim production reached 1,000 aircraft before giving way to the infinitely more powerful Manchester as new capacity at Chadderton and Woodford brought thousands of new operatives to expand beyond recognition the peace-time labour force. The training of this influx in conditions where the demand for aircraft raced ahead of the growth of available space and where stringent blackout regulations combined with the frequent presence overhead of enemy aircraft brought massive and successful effort by management to combat the problems and even the daylight bombing of Chadderton on Easter Monday, 1941 failed to cause material hindrance to production.

If the Anson was beloved by almost all who flew her, the same could certainly not be said of the Manchester. Despite admirable flying qualities, a good bomb load and eight gun armament in three power-operated turrets, the aircraft was bedevilled by frequent failures of the Rolls Royce Vulture engines and production ceased when only 200 had been built. The proven qualities of the airframe led to efforts to replace the wayward Vultures with other engines of comparable power and the Avro design team under Roy Chadwick evolved the Manchester III with a new centre section and four Rolls Royce Merlins. The proposal for the Manchester III was unpopular with the Ministry of Aircraft Production at the time, because every Merlin engine was needed for the fighter

The first Metrovick built Manchester R5768. This aircraft was lost along with twelve others when the Trafford Park factory was bombed on 23rd December 1940. *:G.E.C. Switchgear Ltd*

defence of the UK. However, with the assistance of Mr E (later Lord) Hives of Rolls Royce, Merlin engines were obtained and the first aircraft BT408 was put together in the hangar taken over at Ringway. The Company's Chief Test Pilot, H.A. Brown made the first flight on January 9th 1941.

The flying qualities of the Manchester design and the proven reliability of the Merlin blended to make an aircraft which was initially and continuingly successful in every possible way. The name 'Lancaster' was adopted and production was immediately put in hand on the largest possible scale, the first production aircraft, L7527, flying at Woodford on the 31st October 1941. Whilst its flying and fighting qualities are well renowned, the making of its success stemmed also from another fundamental design attribute — it was easy to make. M.A.P. experts had been unable to accept Dobson's and Chadwick's prediction of 80 aircraft per month (this coming from a firm not used to producing large aircraft), but the achieved rate of 150 per month from Chadderton and Woodford alone was due to the ease of manufacture. Less than a third of the productive labour needed to be skilled tradesmen, the major components were fully fitted with all their systems before final assembly. So the Lancaster was built with fewer man hours per aircraft than any type of comparable size. The organisation to attain this result and support other Lancaster manufacturers with drawings and parts, whilst Anson production still continued apace was truly remarkable.

Later the Lancaster II with Bristol Hercules engines was to be produced in numbers and American built Packard Merlins were to be used in large numbers as production built up, the aircraft flying from Ringway and Woodford as well as from additional factories at Yeadon and in Canada. In addition to the factories directly responsible for Lancaster production, many smaller works in the Manchester area were involved in sub-contract work and the overall work force involved thousands of local people as output built up to an assembly rate of ten aircraft per day for Bomber Command. The use to which Bomber Command put the Lancaster is a matter of National history and the subject of several well-documented books. Probably the best publicised single raid was that of 617 Squadron — the Dam Busters — which blended the expertise of the crews, the ingenuity of Barnes Wallis and the skills of Manchester engineers to bring about the destruction of the Möhne and Eder dams on the night of the 16th/17th May, 1943 after the Squadron's aircraft had been specially modified to take Wallis's spinning mine. Bombs of up to 22,000 lb were eventually carried as the Lancasters' loads increased and Sir Arthur Harris who commanded the bombing force paid tribute to Dobson, Chadwick and 'the thousands of men and women who laboured with skill and devotion' in praising the aircraft as a major factor in the defeat of the Nazi enemy.

Lancaster production took pride of place in the Avro effort but did not prevent other work being carried on. American Douglas Boston and Martin Maryland aircraft for the Royal Air Force were modified in 1941/42 to bring them up to Royal Air Force standards and, as the need for transports became apparent, Roy Chadwick designed the Avro York using the proven Lancaster wing and creating a high-wing aircraft which flew at Ringway on the 5th July 1942, less than six months after commencement of construction at Chadderton. Over 200 Yorks were built and production aircraft were reaching the RAF as well as entering service with British Overseas Airways Corporation as the war entered its later stages. Design capacity was also devoted to a more advanced civil transport for post-war use and the first Tudor was approaching completion as the war ended.

Metrovick built Lancaster BI 'PO-S' R5868; S-Sugar of 467 Squadron. This aeroplane which completed 137 operations over enemy territory is now preserved in the RAF Museum at Hendon. *:G.E.C. Switchgear Ltd*

The Lancaster was adapted to produce the Lancastrian transport for both Service and civil use and with the prospect of redirection of major RAF bombing efforts to bring about the defeat of Japan, the Avro Lincoln was designed to a 1943 specification. The first Lincoln, PW925, flew at Ringway on the 9th June, 1944 and there were plans for large scale use of the type in operations in the Pacific but the hey-day of the Lincoln came later when the Royal Air Force was re-moulded to its peace-time form.

Even before the start of the war it had become apparent that, as well as the facilities of the established aircraft manufacturers, additional manufacturing capacity would be required if the expansion of the Royal Air Force was to be carried on at the required rate. To supplement the Avro output of Manchester bombers, the Metrovick complex at Trafford Park was selected and by November 1940 the first aircraft was ready for assembly. On the 23rd December a direct hit destroyed this aircraft and twelve more but, although production was delayed, a total of 43 Manchesters left the Mosley Road Works and the Manchester was followed on the line by the Lancaster. Before the war ended 1,080 Lancasters had been produced at Trafford Park, including the famous R5868, *S for Sugar* which completed 137 operations and can today be seen in the Royal Air Force Museum. Production of Lincolns was later undertaken and, in addition, Metrovicks produced both Messier and Dowty undercarriages. Automatic pilots were turned out in large numbers and the Company took a major part in work on the radar installations which were to play a vital part in both defensive and offensive air operations. Whilst this work was going on, a revolutionary form of aircraft propulsive unit was developed in conditions of considerable secrecy and on the 29th June, 1943 a Metrovick F2 jet engine was air-tested for the first time in a converted Lancaster. On the 13th November, 1943 the first flight of a British aircraft propelled by axial-compressor type jet engines — a specially modified Gloster F9/40 with two similar engines — took place at Farnborough.

Not far from the Metrovick works, the Trafford Park factory of F. Hills & Sons Ltd commenced its adjustment to war by producing quantities of aircraft plywood and wooden propellers. Praga production had ceased and the Company's original hangar at Barton had been requisitioned but contracts for the manufacture of Anson components and, later, complete Percival Proctor aircraft were undertaken. For this work Hills moved into a newly built hangar at Barton but, until this was completed, a Bellman hangar at Ringway was shared with the Air Transport Auxiliary whose Captain A.W. Vincent air-tested the first Proctors produced there. Conditions at Ringway were far from satisfactory but, with the move to Barton, the rate of production increased and the Company went on to provide the Royal Air Force with over 700 Proctors for use as radio trainers and communications aircraft. The unsuccessful Hillson Pennine had been followed by a small midwing monoplane powered by a 90 hp Cirrus Minor and named the Helvellyn. Only one, G-AFKT, was built and this served the Company as a communications aircraft in the early years of the war. Its designer, Norman Sykes, then evolved a novel aircraft to test the possibility of increasing the range of fighter aircraft by providing a detachable upper wing to enable take-offs to be undertaken with much larger fuel loads. Known as the Hillson Bi-mono, the aircraft was basically a low-wing monoplane using a de Havilland Gipsy Six engine and, flew both in that form and as a biplane before the first release of the upper wing over the Irish Sea after taking off from Squire's Gate in the hands of Percy Richmond — later to be tragically killed at Barton in a Prentice in 1959. The tests were sufficiently successful to justify the modification and testing of a Hurricane as the Hillson FH40 with a similarly detachable upper wing but, although tests were made at Sealand and Farnborough the project was not pursued.

At Heaton Chapel the Fairey Aviation Co Ltd commenced the war with the Battle single-engined bomber in full production and it was a Battle from that factory which scored Britain's first 'kill' of the war when the gunner of K9293 shot down a Messerschmitt Me109 over Northern France. The Battle was, however, too vulnerable to survive in the company of the fighters of the day and was diverted to the training role with several aircraft giving stalwart service as flying test-beds for a variety of engines. 1,171 were built at Stockport, including 16 supplied to the Belgian Air Force, before production ceased in 1942 by which time the fleet fighter Fairey Fulmar was on the assembly lines. The first Fulmar, N1854, flew at Ringway on the 4th January, 1940 after the design had been evolved from the single Fairey P4/34 by Heaton Chapel design staff and of over 600 of the type built, that aircraft is today the only survivor, having been retained by Faireys and subsequently placed in the Fleet Air Arm Museum at Yeovilton. Whilst Fulmar production was building up, the factory received and reconditioned 100 Swordfish biplanes.

From the initial factory and a single small hangar at Ringway, the Fairey Aviation Co rapidly expanded its northern organisation. New assembly sheds were built at Ringway, the Burtonwood Repair Depot was managed for the Ministry of Aircraft Production until the United States Army Air Force took over in 1943 and ancillary and repair plants established including works at Warrington, Stretton, Reddish and the requisitioned bus depot at Parrs Wood. In addition, the Crossley

Fairey Barracuda DP855/G which was converted to the TRIII prototype, seen at Ringway in February 1944. *:Crown Copyright*

Motors works between the original Fairey factory and the Manchester to Stockport railway line was taken over as a Government 'shadow factory' and, under Fairey management, commenced construction of Bristol Beaufighter night fighters. The first Beaufighter left the lines in January 1941 and 500 were built before the Errwood Park Works moved on to the building of Handley Page Halifax heavy bombers. Despite the pre-eminence of the Lancaster, it should not be forgotten that the Halifax made a major contribution to the effectiveness of Bomber Command's operations and over 660 were delivered to the Royal Air Force after test flying at Ringway.

In the later years of the war the Heaton Chapel factory was responsible for providing the Fleet Air Arm with over 1,000 Barracuda torpedo/dive bombers. Ungainly in appearance, the Barracuda carried a wide variety of external stores for delivery to the enemy and, as well as carrying out many anti-submarine patrols from escort carriers, the type was used for wide-ranging anti-shipping attacks, the most spectacular being a successful raid on the German battleship *Tirpitz* on the 3rd April, 1944. Barracudas also carried the war to the Japanese from the aircraft carriers of the British Pacific Fleet.

On the 8th May, 1945 Winston Churchill announced in the House of Commons the unconditional surrender of Germany. On the 6th August the first atomic bomb exploded over Hiroshima and by midnight on the 14th August Britain's new Prime Minister, Clement Attlee was able to announce that the war was effectively over.

After the Holocaust

In the thirty-one years which have passed since the end of the war, it has become almost impossible to pick up a local newspaper without finding reference to some aspect of aviation in relation to the Manchester area. To chronicle the events of the period in any detail would require a large volume and a true assessment of their relative importance is probably for a later generation with the advantage of a longer view and a broader perspective.

The scope of this booklet is not such as to permit more than a brief scanning of the scene but Manchester has seen and is seeing great work in all the many facets of aviation and perhaps a later work will set out to record more fully what has transpired in this still current period.

As successive governments have truncated the nation's forces, the area's direct relationship with the Royal Air Force has faded. No 1 P.T.S. left Ringway for Abingdon in 1946 and, after an enthusiastic post-war existence with first Spitfires and then Vampires, 613 (City of Manchester) Squadron disbanded along with the other Squadrons of the Royal Auxiliary Air Force in 1957 after its Honorary Air Commodore, Sir Roy Dobson, had received from the Lord Mayor an illustrated scroll expressing the City's appreciation of the work of the Squadron. The ceremony took place in Albert Square as a formation of the Squadron's Vampires flew over in salute. The Manchester University Air Squadron is now at Woodvale, near Southport, flying Scottish Aviation Bulldogs after many years with de Havilland Chipmunks and the RAF is represented in the area by Careers Information Offices at Barnett House in Manchester and also in the surrounding towns. The Royal Observer Corps stood down at the end of the war but was reactivated in 1947, the Manchester operations room remaining until 1961 under the command of Observer Commander J.B. Russell. The role of the Corps progressively changed until the main function is now the monitoring of the effects of nuclear weapons and reorganisation has reduced the number of posts but dedicated volunteers still train at Glossop, Poynton, Prestwich, Altrincham and Atherton.

Although both the number and size of units was drastically reduced when the end of the war cut the requirement for RAF personnel, the Air Training Corps has continued to provide recruits to the Service and has also done great work in a wider social context. Wing Headquarters at Levenshulme and Bury still control a large number of Squadrons in the area, many of which have distinguished themselves in competition with similar units throughout the country. The men and women who went from Manchester to join the flying services have, in many cases, sought to recall comradeship of wartime days and the Royal Air Forces Association has several branches and clubs as well as providing an annual flying display at Woodford. Although reduced in numbers, the proud veterans of the Royal Flying Corps Association still meet together and the White Eagles of the Polish Air Force who have made their homes in the area have their own club at Failsworth.

When aviation first impinged upon Manchester those who pioneered the construction of aircraft as a business venture were joined by others who sought from flying an opportunity of a thrilling or pleasurable activity for their leisure hours. In the years since the Second World War many more have followed a similar course and the Lancashire Aero Club, since recommencing flying as its members returned to civilian life, has grown and thrived at its post-war base at Barton. There have been uneasy times but despite initial difficulties in the years immediately after the war, the efforts of a succession of Chief Flying Instructors — not least the greatly experienced Squadron Leader A.A. (Charlie) Rollo — and the work of enthusiastic volunteers have maintained the club's position as the focal point of private flying in the North West. At Great Hucklow the Derbyshire and Lancashire Gliding Club has built up a similarly eminent position in its field, amongst other notable events staging the World Gliding Championships in 1954. Other gliding enthusiasts have flown regularly at Woodford and in the more recent years exponents of the newer sport of hang-gliding have explored the slopes of the area. The exhibition jumps from early balloons have found their counterparts in the formation of Sky Diving Clubs and, after being absent from the scene for many years the stately balloons have again been seen frequently as the advent of portable gas burners has brought popularity to their use for both sporting and exhibition purposes. The invention of the hovercraft found Manchester's experimenters anxious to develop the vehicle for both work and pleasure and in 1967 a Levenshulme dentist, George Armitage, established a world speed record for light hovercraft with his *Manchester Maid*.

The post war years saw the emergence of a new type of aviation enthusiast, unable to take an active part in construction or flying but attracted to the places where aircraft could be seen. Large numbers of young people spent long hours logging the activities and, with other older men for whom mundane peacetime occupations could not obliterate nostalgia for former days in the flying Services, banded together to form clubs and associations which, as well as recording the current aviation scene, began to seek out and set down the history of aviation's earlier years. Manchester has, perhaps more than any other area of the country, produced large numbers of young enthusiasts for many of whom early interest in aviation has been the precursor of a useful career in aircraft production or in Service or Civil Aviation. Over the years several clubs and societies have been formed, perhaps the most notable being the Ringway Spotters' Club which, after passing through a number of phases as its members matured, developed into the present day Manchester Aviation Society. Notable for the variety and calibre of its speakers, the Society has a waiting list for membership and, under the Presidency of Mr J.H. Orrell, holds regular meetings of historians and enthusiasts at the Faculty of Art and Design of Manchester Polytechnic.

The more tangible aspects of aviation history have also found support in the Manchester area. Limited funds and space have prevented the North West Museum of Science and Technology from gathering together as much aviation material as its staff would have wished but a small collection of engines has been established and at Manchester Airport efforts are being made to gather together records and historical momentoes. In 1962 a small band of people formed the Northern Aircraft Preservation Society which has been responsible for the preservation of over thirty aircraft and a similar number of notable engines. From initial use of a lock-up garage at Woodsmoor in Stockport the Society has maintained a position in the forefront of the aircraft preservation movement and, under its new name of The Aeroplane Collection, has plans for the creation of a Museum of Aviation in the North West.

By no means all the Mancunians interested in aviation have been amateur enthusiasts. Manchester's contribution to aviation has taken many forms and has required the efforts of highly trained technicians. The apprentice training given by the major manufacturers and a wide range of courses at the area's technical colleges have been supported by the efforts of both Manchester and Salford Universities. Too little accord has been given publicly to the work which has been done and which continues to be done in this field. Similarly, the work of research which has gone on at the centres of learning is deserving of far wider recognition. The achievements are too numerous to set out in a limited booklet of this nature but it is probably no exaggeration to say that, in the years following the Second World War, the group of scientists at Manchester University under Professors Goldstein and Lighthill contributed more to the development of the theories of high speed aerodynamics than any other single academic group in the world.

Since the war, construction of aircraft has never matched the numerical output of the days of conflict but the relationship between the aviation industry and the region centred upon Manchester has continued with aircraft designed and built locally ranging world-wide and performing a great variety of tasks in both military and civil roles. Changes of emphasis in the Fairey Group over the three post-war decades have seen direct aircraft work reduced until the only visible presence is the work carried out on Fairey Britten-Norman Islanders and Trislanders in the small hangar originally occupied by the Fairey Aviation Co Ltd at Ringway. Fairey Engineering Ltd still provide considerable employment at Heaton Chapel and for some years the factory continued to turn out aircraft for the Royal Navy. As Barracuda production was run down, large numbers of Fairey Firefly general purpose fleet fighters commenced to flow from the factories and the Firefly was succeeded by the Gannet which served the Fleet Air Arm for many years as a search and strike aircraft, continuing to the present day in its airborne early warning role. Large numbers of Fireflash beamriding air to air missiles were also produced at Heaton Chapel and adaptation of Jindivik target drones took place at Ringway. As well as despatching aircraft of Fairey design, the Company's facility at Ringway also dealt with contract work on large numbers of aircraft designed and built elsewhere. Mosquitos, Vampires, Yorks and Invaders all received substantial attention, many of the aircraft subsequently going to overseas air forces.

After wartime production of Manchesters, Lancasters and Lincolns, the giant Metrovick complex ceased to produce airframes, although work continued on the development of jet engines and Hills did not take on any orders to succeed the building of Proctors. The Hawker Siddeley — formerly de Havilland — factory at Lostock which had produced the majority of British wartime propellers progressed to the production of guided missiles of which Firestreak, Red Top, Sea Dart and Martel are the most famous. Ferranti Ltd at Hollinwood and Wythenshawe have continued in the forefront of design and construction of many aids to flying safety and the military effectiveness of Britain's warplanes. Safety equipment has been produced over many years by the Frankenstein Group and many other local firms have, whilst not directly involved in the construction of aircraft, contributed largely in parts and ancillaries.

In recent years Crosby Aviation Ltd at Knutsford have turned out a small run of Andreasson BA-4B aerobatic biplanes and there has been a growing number of enthusiastic constructors who in a variety of unlikely places have undertaken the construction of homebuilt machines under the aegis of the Popular Flying Association. As Fairey production was run down and Avro-designed aircraft were tested at Woodford the large wartime assembly facilities at

Ringway were used principally for storage and refurbishment. After A.V. Roe & Co Ltd left the hangars on the south side of the airfield various aircraft were stored there, including the Junkers JU52/3m transports used briefly by British European Airways Corporation and in the nineteen fifties large numbers of Sabre fighters were refurbished there for the Greek and Turkish Air Forces by Airwork Ltd, Eagle Aircraft Services Ltd later used the site for the major overhaul and rebuilding of Dakotas and Vikings.

Wherever aeroplanes are operated in any numbers, work must be carried on to maintain and repair them and both Barton and Ringway have, in the post-war years, been well served in this respect. Space prohibits a detailed account of the companies specialising in this work but mention must be made of Basil Carlin and Tom Richardson whose company, Barton Moss Engineering Ltd, has performed wonders on aircraft of the Lancashire Aero Club and many other owners over a long period. Servisair Ltd — formerly Manchester Airport Agencies Ltd — has provided engineering services at Ringway in addition to a wide range of handling activities.

Very many companies and people have contributed to aircraft engineering in the area around Manchester, both in construction and maintenance, but, despite the final supercession of the name in the Hawker Siddeley empire in 1963, it is probably true that when the average Manchester citizen thinks of this field, the word 'Avro' will come to mind. After the war Lancasters and Ansons soldiered on in various forms and the Lincoln and York formed for a long time the backbone of the Royal Air Force's Bomber and Transport Commands. The Tudor, bedevilled by political controversy, nevertheless gave useful service, notably in conjunction with the York on the Berlin Air Lift of 1948-49 and on the 9th March, 1949 Jimmy Orrell lifted the first prototype of the Avro Shackleton (VW126) from the runway at Woodford. Powered by four Rolls-Royce Griffon engines, Shackletons have served the Royal Air Force well in the maritime reconnaissance role and will continue to provide airborne early warning cover thirty years after the introduction of the type into service.

To refine the bludgeon of the wartime Bomber Command into the rapier required to provide a deterrent to Britain's potential enemies of the post-war world, Roy Chadwick designed the four-jet Avro 698 using the new and untried delta configuration and, after the commencement of a programme of flying by the small Avro 707 to check the characteristics of the design, the first prototype VX770 flew at Woodford on the 30th August, 1952. Given the name Vulcan by the Royal Air Force, the aircraft still serves as a major part of Britain's reduced deterrent force. Tragically, Chadwick died before his last design took to the air when, on the 27th August, 1947, the Tudor G-AGSU crashed at Woodford and the nation lost one of its truly great aircraft designers.

A small number of Avro Athena advanced trainers were delivered to the RAF and six Ashtons were produced with Rolls-Royce Nene engines for high altitude research purposes. The prototype of an advanced delta winged interceptor, the Avro 720, with combined jet and rocket power was almost complete at Chadderton when the project was cancelled in 1956. Production of Vulcans and Shackletons kept the work force occupied until, in 1960, the first Avro 748 twin Dart low wing transport aircraft was completed under the guidance of a design team led by J.R. Ewans. Over 300 aircraft have so far flown and as the Hawker Siddeley 748 the type continues to be sold for both civil and military purposes. The excellence of the 748 needs no better accolade than its continued use by the Queen's Flight. In production alongside the HS748 is the Hawker

Avro Vulcan B2 of 617 Squadron carrying a Hawker Siddeley Blue Steel missile which was built in quantity at Woodford during the early 1960s.
:Hawker Siddeley Aviation

Siddeley Nimrod which has taken over the Shackleton's role as the RAF's maritime reconnaissance aircraft but with vastly more potent electronic equipment and the power of 4 Rolls Royce Spey jet engines. First flown by Mr John Cunningham in 1967, the Nimrod was designed to use many parts of the Comet 4C and has adequate endurance for long ocean patrols. For shorter range surveillance and with less expensive but still effective search equipment to provide worldwide export potential, a variant of the HS748 is being developed as the Coastguarder and designs for a jet powered version of the 748 are also well advanced.

The beginning of the war saw small numbers of Avro Rota autogiros serving with the Royal Air Force. By the end of the war the helicopter was beginning to enter service and by the time of the Korean conflict when locally built Fairey Fireflies flew from the Royal Navy's aircraft carriers there was no doubt that helicopters would take an ever increasing part in the work of aviation. The years which have followed have seen rotary winged aircraft increasingly used in the Manchester area and their ability to land and take-off away from recognised aerodromes has been evidenced in various parts of the district. Belle Vue and Trafford Park have seen frequent flights and the Police Athletic Ground at Fallowfield has regularly served as a landing place for patients for the Manchester Royal Infirmary. Service helicopters have frequently given demonstrations in parks and on school playing fields and helicopters have supplemented the work of fixed-wing crop-spraying aircraft. For power line inspection and point to point executive transport the use of helicopters has grown to the extent that their presence is now accepted with little comment.

In the field of civil aviation Manchester has in the decades since the war enhanced its standing as a major centre. Over forty years after the gloomy predictions of Captain Smirnoff and despite the encroachment of houses to its borders, Barton continues to provide a home for the Lancashire Aero Club as well as executive and private aircraft and in 1976 the outlook was sufficiently 'bright' to justify the building of a new hangar. Ambitious plans are being made for the provision of an additional aerodrome for light aircraft near to Oldham and at Ringway the aerodrome which forty years ago was envisaged as the centre of a network of services to British towns has grown to become the Manchester International Airport and one of the nation's major embarkation points for Intercontinental flights.

During the war Ringway was not requisitioned but, to provide for the operation of large numbers of heavy aircraft, both newly manufactured and in use by the Royal Air Force, the Air Ministry laid down concrete runways and taxiways. In the years which have followed the development of larger aircraft has brought the need for increased runway length and the main runway, 06/24, has been extended to its present length of 9,200 feet in a number of stages. The small terminal building constructed in 1938 soon proved inadequate as services expanded and the new terminal building opened in 1962 has already been greatly extended with the addition of a third pier to receive the largest aircraft in airline service and considerable extensions of facilities for dealing with the growing numbers of passengers who pass through the Airport.

When, in 1946, the first post-war scheduled service commenced the Dakotas of Air France provided a link between Manchester and Paris. Ringway remained under the control of the City Council but plans for development were delayed by persistent fears that the Minister of Civil Aviation would take over the running of the Airport. Not until 1953 after a prolonged political battle was an agreement reached whereby the Council would retain ownership and the ability to develop the Airport in its own way whilst the Ministry would provide the necessary technical services. Since that time progress has been continuous under municipal control, the present Manchester International Airport Authority having been set up in 1974 to ally the resources of the newly formed Greater Manchester County Council to those of the City. Manchester has been fortunate in its successive Airport Directors, Kenneth Thompson acting in that capacity in the difficult early post-war days after the untimely death of George Lamb and being followed by former master mariner George Harvey, ex-bomber pilot Jack Jackman and the present Director, Gordon Sweetapple, who brought to the position a wealth of experience in aircraft engineering and airport operation.

To detail the participants in the growth of Manchester Airport in the post-war period would require a further book and mention can only be made here of a few. Alongside the scheduled airlines, there have been several small airlines based at varying periods; Sivewright Airways Ltd with a variety of aircraft named after local towns; Melba Airways Ltd and Fingland's Airways Ltd with Rapides and Mercury Airlines Ltd with Dakotas are but a selection from successive local enterprises. Air Views (Manchester) Ltd under the guidance of Bruce Martin and Northern Executive Aviation Ltd whose Managing Director, David Antrobus has taken part in a number of international air races, both continue to provide services to the local business community along with other currently based companies.

Probably the biggest factor in bringing civil aviation into the lives of many local people who might otherwise have had little contact with flying has been the growth of the inclusive tour market whereby relatively inexpensive overseas holidays have become available and Manchester has taken a prominent part in the development of this type of operation, many British and foreign based independent airlines annually taking many thousands of people on holiday from the Airport. Again it is perhaps invidious to select a few from the many operators involved but it is a matter of record that in recent years the aircraft of Dan-Air Services Ltd have not only been operated but maintained at Ringway. Prominent, too, have been the aircraft of Wardair (Canada) Ltd, the first airline to operate regular flights to and from Manchester with the massive Boeing 747 Jumbo Jet.

Regular scheduled air services have expanded massively since the end of the War and continued municipal enterprise has ensured that Manchester has maintained its place as the country's major centre outside London for both passenger and freight flights to the Continent. Alongside services by most of Europe's national airlines, transatlantic services have also been developed. From limited flights by Rapides in 1946 British European Airways operated Dakotas, Ambassadors, Viscounts, Vanguards and Tridents before basing their entire BAC 1-11 fleet as Manchester's own airline. British Overseas Airways Corporation were not the first to operate transatlantic services from Manchester but, once started, expanded their services with successive use of Stratocruisers, DC7s and Britannias, to the regular VC10s and Boeing 707s which have continued in use with British Airways. Shortly after the publication of this booklet the Boeing 747 will take its regular place on services from Manchester to America and already the designation of Manchester International Airport as the main diversion airport for Heathrow has brought the debut of the supersonic Concorde, the first aircraft of the type arriving on the night of the 14th November, 1976.

From the day when James Sadler first ascended from the rear of a gentleman's residence in Long Millgate, Manchester and aviation have both advanced immeasurably in very many ways. To a considerable extent they have done so together. If this booklet has illustrated some of the ways in which this has been so, it will have served its purpose.

The first Concorde to land at Ringway was the British Airways machine G-BOAA shown which was diverted to Manchester on 14th November 1976 — A foretaste of the future. *:Manchester International Airport Authority*

The Manchester Branch of the Royal Aeronautical Society 1927 — 1977

B.A.G. Meads and J.A.E. Waterfall

The Royal Aeronautical Society was founded in 1866 to foster the growing practical and academic interest in the study and development of aeronautics. The first of the extant Branches was formed in Coventry in 1925.

After some early activity in the area in 1919, the Manchester Branch began formally as a branch of the Institute of Aeronautical Engineers. The Charter of Authority to form it was granted to Capt T. Patterson on 10th December 1925, signed by Col L.T.C. Moore-Brabazon amonst others. Over a year was to elapse before the inaugural meeting. Then, on 21st January 1927, at the New Shades Restaurant Mr R.H. Dobson was elected Chairman, Mr A Dennison-Scarlett Secretary/Treasurer and Messrs H. Bolsover, C.G. Carlisle, G.W. Durbin and C.E. Fielding formed the Committee.

Membership was quite small to start with but after the I.Ae.E. was incorporated with the Royal Aeronautical Society in March 1929 the I.Ae.E. branch was disbanded and the R.Ae.S. branch was formed at a meeting on 7th February 1930 attended by the Society's Secretary Capt. J. Laurence Pritchard. A fortnight later, another inaugural meeting was held, this time at A.V. Roe & Co where Mr J.D. Siddeley was elected President, Mr R. Chadwick Chairman, Mr F. Paxton Secretary and Mr A. Dennison-Scarlett Treasurer with a Committee of eight. The new formation was an immediate success and Membership grew to 153 in the first year. The objectives have remained almost unchanged over the years, and the subscription has only doubled from 5 shillings in 1930 to 50p. today.

The interpretation of the objectives was certainly more varied. It was a declared objective to form a library, but in addition to the lectures and visits familiar today, a gliding section and a model aircraft section were formed. Within two years the gliding section had built and flown its own Zögling glider, and had absorbed the Stockport Gliding Club with its damaged Prufling glider. Under the Chairmanship of Basil Meads, the section went on to lay the foundations of a two seater training scheme which was far ahead of its time. A two seater BAC VII glider had been designed by Lowe-Wylde and major components for one were purchased from him at a price of £53. The section set to work to build the BAC VII, making all the other components themselves. At this stage Roy Chadwick was to prove of great value to the project with his help and encouragement. However, when it was completed at Woodford he tempered any over-enthusiasm by making the prospective pilots wait a fortnight whilst the Technical Report was completed and a British Gliding Association Certificate of Airworthiness obtained.

The training programme developed on the BAC VII by the section was recognised firstly in the Slingsby Falcon III two-seater design in 1935 and finally in 1948 by a British Gliding Association ban on solo training.

Meanwhile the model section made its own way, quietly at first, but doing its own merging with the Lancashire Model Aircraft Society. In 1934 Mr C.S. Rushbrooke was Secretary of this section and reported to the AGM that it was seventy strong and had competition successes including the All-England